Logistics and Supply Chain 2.0: The Digital Transformation Guide

Embracing Innovation in Logistics and Supply Chain Management

Ms. Harshita Gaikwad

Dr. Bijal Zaveri

Dr. Varsha Bhabad

Mr. Santosh Gaikwad

To my students,

As we present the second edition of our book, *Logistics and Supply Chain 2.0: The Digital Transformation Guide*, I extend my deepest gratitude to the students, colleagues, and all the supportive members who have played a crucial role in this journey.

To my students, Co-author —your curiosity, enthusiasm, and thirst for knowledge have been a constant source of inspiration. Your thought-provoking discussions and keen interest in the evolving world of logistics and supply chain management have fuelled the insights in this book.

Special thanks to the Dean Faculty of Management Studies, Parul University Dr. Bijal Zaveri Madam, and My Ph.D. Guide Dr.Varsha Bhabad Madam from the Institute of Management and Research, Nashik for constant guidance and support,Professor & Academic Administrative Head of the Faculty of Management Studies, Mr. Santosh Gaikwad, for their unwavering support and guidance.

To my colleagues, mentors, and well-wishers—your encouragement, feedback, and shared passion for advancing this field have been invaluable.Your support has strengthened my commitment to delivering knowledge that is both relevant and impactful.

This book is a testament to the collaborative spirit that drives learning and innovation. I am grateful for your contributions and look forward to continuing this journey of exploration and growth together.

Thank you for being part of this endeavor!

CONTENTS

PREFACE

The rapid evolution of technology has ushered in a transformative era for logistics and supply chain management. As businesses grapple with the complexities of global trade, sustainability, and customer expectations, the role of technology has become pivotal in driving efficiency, transparency, and innovation. It is with great enthusiasm that we present the second edition of our book, *Logistics and Supply Chain 2.0: The Digital Transformation Guide*. This edition builds on the foundation laid by its predecessor, *A Revolutionary Plunge into Logistics & Supply Chain: From Basics to Advance*, and ventures further into the dynamic landscape of digital transformation.

The first edition aimed to provide readers with a comprehensive understanding of logistics and supply chain management principles, covering fundamental concepts and advanced strategies. It was warmly received by practitioners, academicians, and students alike, who appreciated its structured approach and practical insights. Encouraged by this response, we embarked on creating a second edition that not only reflects the technological advancements shaping the industry but also anticipates future trends.

This edition introduces readers to modern technologies that are redefining logistics and supply chain management. Each chapter delves into a critical aspect of digital transformation, offering theoretical insights and practical applications. Topics such as the Internet of Things (IoT), cloud computing, and artificial intelligence explore how these technologies optimize operations and enable real-time decision-making. Emerging tools like augmented and virtual reality, drones, and autonomous vehicles are discussed as game-changers in logistics training and transportation. Sustainability, a pressing global concern, is addressed through chapters on green logistics and sustainable technologies, emphasizing the shift towards eco-friendly practices.

Recognizing the significance of security and transparency in an interconnected world, we have dedicated chapters to cybersecurity in supply chain networks and blockchain technology. These topics underscore the importance of safeguarding data and ensuring trust among stakeholders. Additionally, the growing influence of e-commerce and last-mile delivery innovations is explored, providing a comprehensive view of how consumer behavior drives operational shifts.

The second edition is the result of extensive research, collaboration with industry experts, and feedback from our readers. It is designed to serve as a guide for students, professionals, and organizations aiming to navigate the digital transformation journey in logistics and supply chain management. The practical examples and case studies included in this edition further bridge the gap between theory and practice, making it an indispensable resource for aspiring leaders in the field.

We extend our heartfelt gratitude to our readers, whose encouragement and constructive feedback inspired us to undertake this endeavor. We also thank the contributors and reviewers who provided invaluable insights to ensure the accuracy and relevance of this edition We are also deeply grateful to the co-authoring students for their dedication and efforts in shaping this book.

As the logistics and supply chain industry continues to evolve, we hope this book serves as a trusted companion, equipping readers with the knowledge and tools needed to thrive in a technology-driven world. Let us embark on this journey of innovation and sustainability together, shaping the future of logistics and supply chain management.

– Asst. Prof. Harshita Santosh Gaikwad

01

INTRODUCTION TO MODERN TECHNOLOGY IN LOGISTICS

About the Author - Mr. Shivain Pathela is an MBA candidate specializing in logistics and supply chain management at Parul University, Vadodara. He earned his Bachelor's degree in Business Administration (B.Com) with a specialization in E-commerce from Chandigarh University, Mohali. His academic foundation is complemented by internships at Satia Industries Ltd.—one of India's top 10 paper manufacturers—  Ventura Securities Ltd., and HDFC Bank, where he gained practical experience in sales and various business functions.

A passionate athlete, Shivain has represented at the national level in ball badminton, the district level in basketball, and the zonal level in swimming. These achievements have fostered his leadership, teamwork, and strategic abilities, which further strengthen his professional acumen.

With a blend of academic rigor, practical exposure, and sports discipline, Shivain Pathela is an emerging voice in business literature. His dedication to excellence and insightful perspective on logistics and supply chain management make his contributions both meaningful and impactful.

Introduction to Modern Technology in Logistics

1.1 Introduction

Logistics, the backbone of global trade and supply chain operations, has undergone a remarkable transformation with the advent of modern technology. Traditional logistics systems, often characterized by manual processes and limited data integration, have been replaced or enhanced by innovative solutions that leverage advanced technologies to optimize operations, improve efficiency, and enhance customer experiences.

Modern technologies like blockchain, robotics, artificial intelligence (AI), the Internet of Things (IoT), and big data analytics have completely changed how items are tracked, managed, and delivered within supply chains. Real-time visibility, predictive analytics, and smooth communication are made possible by these developments, which assist businesses in making wise decisions and adjusting to the ever-changing needs of the market.

In an era where speed, accuracy, and cost-efficiency are critical, modern technology has become indispensable in logistics. From automated warehouses and drone deliveries to smart inventory systems and blockchain-powered transparency, these innovations are reshaping the industry, driving competitiveness, and paving the way for sustainable logistics practices.

1.2 Definitions and Meaning

Definition: Modern technology in logistics refers to the integration of advanced tools, systems, and innovations, such as artificial intelligence (AI), the Internet of Things (IoT), blockchain, and robotics, to enhance the efficiency and effectiveness of supply chain operations. It enables real-time tracking, predictive analytics, and seamless communication across various stages of logistics. These technologies streamline processes, reduce costs, and improve accuracy while meeting the increasing demands for speed and sustainability in the global supply chain.

According to Dr. Donald Bowersox, a pioneer in supply chain management, modern technology in logistics involves using advanced systems and tools to ensure seamless integration, automation, and visibility across the supply chain. This includes technologies like artificial intelligence (AI), the Internet of Things (IoT), and blockchain, which enhance decision-making, optimize operations, and improve customer satisfaction.

Similarly, John Gattorna, a renowned logistics expert, emphasizes that modern technology revolutionizes traditional logistics by leveraging data-driven solutions and automation to create responsive and efficient supply chain networks. These technologies not only drive operational excellence but also foster innovation and sustainability in the industry.

Meaning: Modern technology in logistics refers to the use of advanced tools and systems—such as artificial intelligence (AI), the Internet of Things (IoT), robotics, and blockchain—to improve and streamline the movement and management of goods within supply chains. These technologies enhance operational efficiency by automating tasks, providing real-time tracking, improving decision-making, and offering greater visibility across logistics processes. By incorporating these innovations, businesses can reduce costs, increase speed, and meet customer demands more effectively.

1.3 Key Components of Modern Technology in Logistics

- **Artificial Intelligence (AI):** AI helps in automating decision-making processes, optimizing routes, forecasting demand, and improving customer service. It analyzes vast amounts of data to provide insights that drive efficiency and cost savings in logistics operations.

- **Internet of Things (IoT):** IoT involves connecting devices through the internet to gather and share data. In logistics, IoT is used for real-time tracking of shipments, monitoring vehicle conditions, and improving inventory management, leading to better control and visibility across the supply chain.

- **Blockchain:** Blockchain provides secure, transparent, and immutable record-keeping, which enhances traceability and reduces fraud. It is particularly useful in tracking goods across global supply chains and ensuring data integrity, which builds trust among all stakeholders.

- **Robotics and Automation**: Robotics and automation streamline warehouse operations, such as picking, sorting, and packing goods. Automated systems reduce human error, increase speed, and lower labor costs, enabling warehouses to function more efficiently.

- **Big Data and Analytics:** Big data analytics allows logistics companies to analyze vast datasets to improve forecasting, demand planning, and operational performance. It helps in identifying trends, predicting disruptions, and making informed decisions for better supply chain management.

- **Cloud Computing:** Cloud-based platforms allow logistics companies to store and access data remotely, improving collaboration and communication among different supply chain stakeholders. Cloud computing offers scalability, flexibility, and real-time data sharing, enabling more efficient management of logistics operations.

- **Augmented Reality (AR) and Virtual Reality (VR):** AR and VR technologies are increasingly used in logistics for training, warehouse management, and product handling. For example, AR helps warehouse workers locate items quickly by overlaying directions on their view, while VR is used for simulating logistics scenarios for better planning.

- **5G Technology:** The roll-out of 5G networks enhances connectivity and enables faster communication between devices in logistics operations. This technology improves the reliability and speed of real-time data transmission, which is crucial for managing global supply chains and autonomous systems.

1.4 Core Concepts of Modern Technology in Logistics

The core concepts of modern technology in logistics revolve around using advanced technologies to enhance efficiency, accuracy, and sustainability within the supply chain. Here are the key concepts:

1. **Automation and Robotics:** The use of automated systems and robots to streamline operations like inventory management, order picking, sorting, and packing. This reduces manual labor, enhances speed, and minimizes human error.

2. **Real-time Data and Tracking**: The ability to track goods and assets in real time using technologies like IoT and GPS. This

enables better visibility across the supply chain, allowing for proactive decision-making, accurate delivery estimates, and more responsive customer service.

3. **Data Analytics and Artificial Intelligence (AI):** Leveraging big data, AI, and machine learning to analyze vast amounts of data for predictive insights. This includes demand forecasting, route optimization, and risk management, which help companies optimize their operations and improve decision-making.

4. **Supply Chain Transparency and Blockchain**: Using blockchain technology to create a transparent, secure, and immutable record of transactions within the supply chain. This ensures traceability, enhances security, reduces fraud, and builds trust among stakeholders.

5. **Connectivity and the Internet of Things (IoT)**: IoT enables the interconnection of devices, sensors, and systems to collect and exchange data, improving the monitoring and management of goods, inventory, and vehicles in the logistics process.

6. **Sustainability:** Integrating technologies to create more eco-friendly logistics solutions. This includes optimizing routes to reduce fuel consumption, using electric vehicles, and implementing sustainable packaging and waste management practices.

7. **Smart Warehousing and Inventory Management:** The use of advanced technologies to enhance warehouse operations, including automated storage and retrieval systems (ASRS), drone-assisted inventory checks, and real-time tracking of goods to ensure better stock control and faster processing times.

8. **Last-Mile Delivery Solutions:** Leveraging drones, autonomous vehicles, and AI to optimize last-mile delivery, improving efficiency, reducing costs, and ensuring faster deliveries to customers.

9. **Cloud Computing:** Centralizing logistics data and systems in the cloud to enable real-time collaboration and access to information across the supply chain, leading to improved coordination and decision-making.

1.5 Here are some additional core concepts of modern technology in logistics:

1. **Digital Twin Technology:** This involves creating a virtual replica of physical logistics assets (like warehouses, vehicles, or entire supply chains) to simulate, monitor, and optimize operations. Digital twins help predict potential issues and optimize performance in real time.

2. **Predictive Analytics:** Using historical data and AI models to predict future trends, demand fluctuations, and supply chain disruptions. This helps businesses proactively adjust their operations to minimize risks and optimize resources.

3. **Supply Chain as a Service (SCaaS):** This concept involves leveraging cloud-based solutions to outsource logistics services, including transportation, warehousing, and inventory management, providing scalability and flexibility without the need for heavy upfront investment.

4. **Electronic Data Interchange (EDI):** The use of standardized electronic systems to exchange business documents (like invoices, orders, and shipment notices) between companies. This speeds up processes, reduces errors, and ensures smoother communication between stakeholders.

5. **Smart Packaging:** Incorporating sensors and technologies like QR codes or RFID into packaging to track goods and ensure the integrity of products, such as monitoring temperature-sensitive items during transportation.

6. **Advanced Fleet Management: Utilizing telematics,** IoT sensors, and AI-driven software to monitor, manage, and optimize the performance of transportation fleets. This includes route optimization, fuel consumption monitoring, and predictive maintenance.

7. **Crowdsourced Delivery:** Leveraging gig economy platforms and crowd-sourced delivery methods to optimize last-mile

logistics, allowing businesses to tap into a wider pool of delivery resources and improve delivery speed and efficiency.

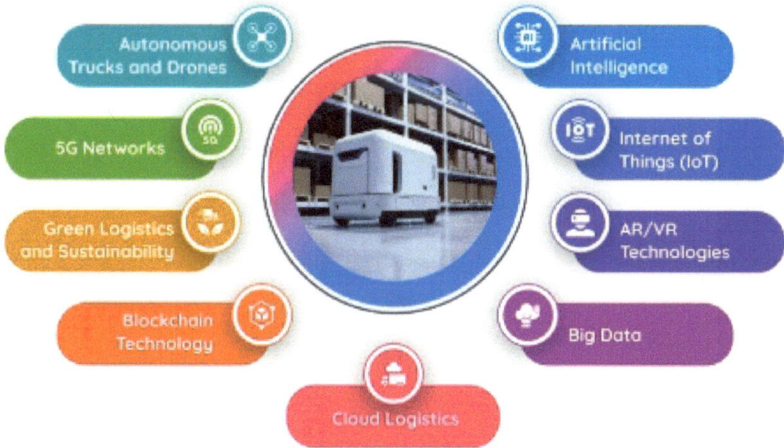

1.6 Difference Between Traditional And Modern Technology In Logistics

Aspect	Traditional Technology	Modern Technology
Communication	Manual communication via phone, fax, or paper	Digital communication via email, apps, and IoT
Data Management	Paper-based records	Cloud-based systems and digital databases
Tracking and Monitoring	Limited to manual checks and physical inspection	Real-time GPS tracking and IoT-enabled devices
Inventory Management	Manual counting and static records	Automated systems with AI-driven analytics
Efficiency	Time-intensive and prone to errors	Streamlined processes with automation

Aspect	Traditional Technology	Modern Technology
Cost	Higher operational costs due to inefficiencies	Reduced costs through optimized resource usage
Customer Interaction	Limited to basic communication methods	Advanced CRM tools and personalized experiences
Speed of Operations	Slower due to manual processes	Faster with automation and predictive analytics
Adaptability	Difficult to adjust to changes quickly	Flexible and easily scalable systems
Sustainability	Often resource-intensive with waste	Focus on eco-friendly and optimized solutions
Decision-Making	Based on experience and intuition	Data-driven decision-making with AI and ML
Supply Chain Visibility	Fragmented and limited	End-to-end transparency using advanced tech
Integration	Siloed systems with little interoperability	Integrated platforms for seamless operation

1.7 Challenges and Impacts of Modern Technology in Logistics

Modern technology in logistics has significantly transformed the industry, but it also comes with various challenges. Here are the key ones:

1. Integration and Compatibility

Challenge: Many companies use legacy systems that are not compatible with modern technologies like IoT, AI, and blockchain.

Impact: Inefficiencies in data sharing and decision-making across supply chain networks.

2. Cybersecurity Risks

Challenge: Increased use of digital platforms and IoT devices makes logistics systems vulnerable to cyberattacks.

Impact: Data breaches can lead to financial losses, operational disruptions, and reputational damage.

3. High Implementation Costs

Challenge: Adopting advanced technologies like autonomous vehicles, drones, or robotics often requires significant investment.

Impact: Smaller companies may struggle to compete with larger players who can afford these innovations.

4. Data Management Issues

Challenge: The logistics sector generates massive amounts of data, and managing, analyzing, and securing it effectively is complex.

Impact: Poor data management can lead to inefficiencies, missed opportunities, and compliance issues.

5. Real-Time Visibility

Challenge: Achieving full visibility across the supply chain is still difficult due to fragmented systems and lack of standardized tracking methods.

Impact: Delayed or inaccurate information affects delivery schedules and customer satisfaction.

6. Talent Shortage

Challenge: A lack of skilled workers to operate and maintain advanced technologies.

Impact: Limits the ability of companies to fully leverage innovations in logistics.

7. Sustainability Pressures

Challenge: Governments and consumers are demanding greener logistics operations, but implementing sustainable practices can be expensive.

Impact: Balancing cost-efficiency with eco-friendly practices remains a major hurdle.

8. Regulatory Compliance

Challenge: Navigating complex and evolving global trade regulations, especially in cross-border logistics.

Impact: Non-compliance can lead to penalties, delays, and loss of business.

9. Scalability Issues

Challenge: Ensuring that technology systems can scale effectively to meet fluctuating demand or expand to new markets.

Impact: Poor scalability can hinder growth and responsiveness.

10. Dependence on Technology

Challenge: Over-reliance on technology can lead to vulnerabilities during system outages or technical failures.

Impact: Downtime can disrupt operations and erode trust with customers.

Addressing these challenges requires strategic investment, a focus on collaboration, and a commitment to continuous improvement in both technology and processes.

1.8 Overcoming Form Challenges Faced in Modern Technology in Logistics

To overcome these challenges in logistics caused by modern technology, organizations need a comprehensive and strategic approach that integrates technical solutions, workforce development, and industry-wide collaboration. Here's how to address each challenge:

1. Integration and Compatibility

- **Adopt Middleware Solutions:** Implement integration platforms that enable seamless communication between legacy systems and modern technologies like IoT and blockchain.
- **Phased System Upgrades:** Gradually replace outdated systems with modern, compatible ones to ensure continuity while minimizing disruptions.
- **Standardization:** Collaborate with industry players to develop and adopt standardized data-sharing protocols.

2. Cybersecurity Risks

Robust Security Protocols: Invest in advanced cybersecurity measures such as encryption, firewalls, and intrusion detection systems.

Regular Audits and Updates: Conduct routine system audits and update software to patch vulnerabilities.

Employee Training: Train staff on recognizing and preventing cyber threats like phishing attacks.

3. High Implementation Costs

Leverage Partnerships: Partner with technology providers to share costs and access flexible financing options.

Pilot Programs: Test new technologies in smaller segments before full-scale implementation to minimize risks and costs.

Focus on ROI: Prioritize investments in technologies with clear, measurable cost savings and operational efficiencies.

4. Data Management Issues

Adopt Advanced Analytics Tools: Use AI and machine learning to process and analyze large data volumes for actionable insights.

Cloud-Based Platforms: Shift to cloud solutions for scalable, secure, and cost-effective data storage and management.

Data Governance Framework: Implement policies to ensure data quality, security, and compliance with regulations.

5. Real-Time Visibility

IoT and GPS Tracking: Deploy IoT devices and GPS for real-time tracking of shipments and assets.

Integrated Dashboards: Use unified dashboards to consolidate and display data from various systems.

Blockchain Technology: Utilize blockchain for transparent and tamper-proof supply chain data.

6. Talent Shortage

Upskilling Programs: Invest in training employees to operate and maintain advanced logistics technologies.

Collaborations with Academia: Partner with educational institutions to develop specialized courses in logistics and technology.

Automate Repetitive Tasks: Use automation to reduce dependency on manual labor for routine processes.

7. Sustainability Pressures

Eco-Friendly Technologies: Invest in green technologies like electric vehicles, renewable energy, and carbon-neutral shipping options.

Efficient Route Planning: Use AI for optimizing routes to reduce fuel consumption.

Sustainability Certifications: Work towards certifications that demonstrate compliance with environmental standards.

8. Regulatory Compliance

Regulation Monitoring Tools: Use automated tools to track and adapt to changing regulations in real-time.

Expert Consultation: Engage trade compliance experts to navigate complex global regulations.

End-to-End Visibility: Ensure documentation and tracking systems are in place for all stages of the supply chain.

Conclusion

In conclusion, modern technology has revolutionized the logistics industry, driving efficiency, transparency, and scalability. Innovations such as IoT, AI, blockchain, and robotics have enabled real-time tracking, data-driven decision-making, and automation of complex processes. However, these advancements come with challenges, including high implementation costs, cybersecurity risks, integration issues, and the need for skilled talent. Companies must strategically address these hurdles by investing in robust systems, enhancing cybersecurity, and fostering industry collaboration.

Sustainability and compliance with evolving regulations remain critical priorities, urging firms to balance technological growth with environmental responsibility. Despite the obstacles, the integration of modern technologies offers immense potential for improving customer satisfaction, reducing operational costs, and achieving global supply chain resilience. By continuously adapting and embracing innovation, logistics companies can ensure long-term success in an increasingly interconnected world.

02

THE TECHNOLOGY-DRIVEN SHIFT TO SUSTAINABLE SUPPLY CHAINS

Author's Bio – Mr. Adnan Mahedi Panwala is currently pursuing an MBA with a specialization in Logistics and Supply Chain Management at Parul Institute of Management and Research, Vadodara, following a Bachelor of Science degree with Chemistry as his core subject from Govind Guru University, Gujarat. His solid scientific background equips him with an analytical mindset that complements his understanding of supply chain dynamics. During his internship at **Polycab India Limited**, Halol, Gujarat, he gained significant industry exposure, applying his problem-solving and creative thinking skills to optimize logistics processes and enhance supply chain efficiency. This experience provided him with hands-on insights into the practical challenges of supply chain management.

In addition to his internship, Adnan is actively engaged in academic research and he is focused on the analytical aspects of supply chain management.

Adnan's academic journey is complemented by a keen interest in the technological advancements shaping the future of supply chain management. He has completed various courses in data analytics from reputed institutes, developing the expertise needed to leverage data and AI in enhancing supply chain operations. His vision for the future revolves around integrating these technologies to create supply chains that are more resilient, adaptable, and capable of meeting the demands of an increasingly complex global environment.

Introduction to Sustainable Technologies in SCM

The integration of sustainability into supply chain management (SCM) has become an essential priority for businesses in the 21st century. Historically, supply chains were primarily focused on cost efficiency, speed, and reliability, often at the expense of the environment. However, increasing awareness of climate change, stricter environmental regulations, and rising consumer demand for eco-friendly practices have driven a shift toward sustainable supply chains. Sustainable technologies in SCM represent tools and processes designed to reduce environmental impacts, conserve natural resources, and promote social responsibility without compromising operational efficiency or profitability.

India, as one of the fastest-growing economies in the world, plays a critical role in the global supply chain. With initiatives like the **National Electric Mobility Mission Plan (NEMMP)** and **Faster Adoption and Manufacturing of Hybrid and Electric Vehicles (FAME)**, the Indian government is actively promoting the adoption of sustainable technologies in the logistics and transportation sector. Additionally, large Indian corporations such as **Reliance Industries**, **Tata Group**, and **Mahindra & Mahindra** are investing heavily in renewable energy, electric mobility, and circular supply chain models to align with sustainability goals.

The adoption of these technologies not only helps businesses align with global sustainability goals, such as the United Nations' Sustainable Development Goals (SDGs), but also provides long-term economic benefits. Companies like **Amazon India**, **Flipkart**, and **Dabur** have already begun incorporating renewable energy systems, electric delivery vehicles, and digital tools to minimize their environmental footprint. These efforts highlight how sustainable technologies are reshaping the supply chain landscape in India, paving the way for a more resilient and responsible future.

The Growing Importance of Sustainability in SCM

The supply chain is responsible for a significant portion of a company's environmental footprint, including energy consumption, carbon emissions, and waste generation. According to recent studies, logistics and transportation alone contribute over **8% of global greenhouse gas emissions**, while manufacturing and warehousing are key contributors to energy consumption and waste. In India, the problem is exacerbated by the heavy reliance on road transportation, which accounts for nearly **65% of freight movement**, contributing to high levels of air pollution and greenhouse gas emissions.

Businesses today face mounting pressure from three key sources:

1. Regulatory Compliance

Governments and international bodies are enforcing stricter environmental regulations, such as the European Union's Green Deal and carbon emission targets. In the Indian context, the government has launched multiple initiatives to encourage sustainability, such as:

- **Perform, Achieve, and Trade (PAT) Scheme:** A market-based mechanism to improve energy efficiency in energy-intensive industries.
- **National Action Plan on Climate Change (NAPCC):** This framework focuses on promoting renewable energy, energy efficiency, and sustainable urban development.
- **Extended Producer Responsibility (EPR):** Under this rule, companies are required to take responsibility for the recycling and disposal of their products, especially in sectors like electronics and plastics.

2. Consumer Demand

Modern consumers are increasingly favoring brands that prioritize sustainability and demonstrate social responsibility. In India, consumer awareness about sustainability is on the rise, particularly among urban populations. A growing number of Indian customers prefer eco-friendly packaging, ethically sourced materials, and products from brands that demonstrate a commitment to environmental protection. Companies like **Tata Consumer Products** and **Godrej** are focusing on reducing plastic usage, adopting biodegradable packaging, and incorporating sustainability into their business models to appeal to eco-conscious consumers.

3. Competitive Advantage

Companies with green supply chains often gain a competitive edge by reducing costs, mitigating risks, and enhancing customer loyalty. In the Indian market, where competition is fierce, sustainability initiatives are becoming a differentiator. For example:

- **Flipkart** has introduced electric delivery vehicles as part of its logistics fleet, aiming to transition to 100% electric by 2030 under the **Climate Group's EV100 initiative**.
- **ITC Limited** has embraced sustainable practices by achieving carbon-positive and water-positive status, showcasing how Indian companies can gain an advantage through sustainability.

As India aims to achieve its **net-zero emissions target by 2070**, businesses that adopt sustainable supply chain practices today will be better positioned to meet future regulatory requirements and market expectations. By leveraging sustainable technologies and addressing India's unique challenges, the nation can lead the way in creating environmentally responsible supply chains while driving economic growth.

Key Sustainable Technologies in SCM

Sustainable technologies span various stages of the supply chain, including transportation, warehousing, packaging, and resource management. Each innovation contributes to reducing waste, improving resource efficiency, and mitigating the environmental impact of supply chain activities. Below is a detailed exploration of the key sustainable technologies transforming SCM.

1. Electric Vehicles (EVs) and Alternative Fuels

Transportation remains one of the largest contributors to greenhouse gas emissions in supply chains, accounting for a significant percentage of global carbon emissions. The introduction of electric vehicles (EVs) and alternative fuels has revolutionized this sector, providing cleaner and more energy-efficient transportation solutions.

- **Electric Vehicles (EVs):**

  EVs are powered by electricity rather than fossil fuels, producing zero tailpipe emissions. They are increasingly being adopted for freight transportation and last-mile delivery, particularly in urban areas where air pollution is a critical issue. Advances in battery technology have improved the range and efficiency of EVs, making them a viable option for long-haul transportation. For instance, **Tesla's electric semi-trucks** are designed to reduce fuel consumption in freight operations. Additionally, governments worldwide are offering incentives and subsidies to encourage businesses to transition to EV fleets, further accelerating their adoption.

- **Alternative Fuels:**

  Alternative fuels such as biodiesel, hydrogen fuel cells, and compressed natural gas (CNG) offer cleaner options for powering vehicles and equipment.

  - **Biodiesel:** A renewable fuel derived from organic materials, biodiesel significantly reduces carbon emissions compared to conventional diesel.
  - **Hydrogen Fuel Cells:** Hydrogen is an emerging fuel source that produces water as its only by-product, making it ideal for heavy-duty vehicles.

- ○ **CNG and LNG:** Compressed natural gas (CNG) and liquefied natural gas (LNG) are widely used in long-haul trucking due to their lower emissions and cost advantages.

2. Energy-Efficient Warehousing

Warehousing plays a critical role in SCM but is also one of the most energy-intensive operations. Sustainable technologies have transformed warehouses into energy-efficient hubs that reduce costs and environmental impact.

- **Smart Lighting Systems:**

  Traditional lighting systems consume large amounts of electricity, often leading to excessive energy use. The introduction of LED lighting combined with motion sensors and daylight integration has significantly reduced energy consumption. LED lights are not only more efficient but also last longer, requiring less frequent replacement and maintenance.

- **Renewable Energy Systems:**

  Warehouses are increasingly integrating renewable energy sources such as solar panels and wind turbines to power operations. These systems reduce reliance on grid electricity and lower carbon emissions. For example, **Amazon's fulfillment centers** have installed solar panels on their rooftops to generate clean energy for day-to-day operations.

- **Smart HVAC Systems:**

  Heating, ventilation, and air conditioning (HVAC) systems in warehouses are being equipped with IoT-enabled sensors that monitor and adjust temperature and airflow based on real-time requirements. This ensures optimal energy use while maintaining suitable working conditions.

- **Green Building Design:**

  Warehouses are now being constructed using eco-friendly materials and incorporating energy-efficient designs. Features like natural ventilation, insulated walls, and green roofs reduce heating and cooling requirements, further minimizing energy consumption.

3. Circular Supply Chains

Traditional supply chains often follow a linear model: materials are extracted, processed, used, and discarded. This approach generates significant waste and depletes natural resources. Circular supply chains, on the other hand, aim to close the loop by reusing, recycling, and repurposing materials.

- **Reverse Logistics:**

  Reverse logistics systems efficiently manage product returns, repairs, and recycling. For example, e-commerce companies like **Zalando** have developed advanced reverse logistics networks to handle high volumes of product returns while ensuring materials are recycled or reused.

- **Recycling Technologies:**

  Modern recycling technologies use robotics and AI to sort and process waste materials, recovering valuable components for reuse. This reduces the need for virgin raw materials and lowers overall production costs.

- **Product Lifecycle Management (PLM):**

  PLM software helps businesses track the lifecycle of their products, from production to disposal. This allows companies to identify opportunities for refurbishing or recycling materials, reducing waste and maximizing resource utilization.

4. Internet of Things (IoT) for Sustainability

The Internet of Things (IoT) has become a cornerstone of sustainable supply chains by enabling real-time monitoring and optimization. IoT devices and sensors are used to collect data on energy use, inventory levels, and environmental conditions, providing actionable insights to improve efficiency.

- **Energy Monitoring Sensors:**

  IoT sensors installed in warehouses and manufacturing facilities track energy consumption and identify areas for improvement. These systems enable businesses to reduce energy waste and set efficiency benchmarks.

- **Cold Chain Management:**

  Temperature-sensitive goods, such as pharmaceuticals and perishable foods, require precise storage and transportation conditions. IoT-enabled temperature sensors ensure these conditions are maintained, reducing spoilage and waste. For example, IoT sensors can send real-time alerts if refrigeration units malfunction, allowing for immediate corrective action.

- **Smart Asset Tracking:**

  IoT devices track the location and condition of goods throughout the supply chain. This ensures transparency, reduces delays, and minimizes inefficiencies caused by lost or damaged shipments.

5. Blockchain for Transparency and Traceability

Blockchain technology is revolutionizing supply chains by providing a secure, transparent, and tamper-proof record of transactions. This fosters accountability and trust among stakeholders while ensuring compliance with sustainability goals.

- **Ethical Sourcing:**

  Blockchain enables companies to verify the origin of raw materials and ensure they are sourced ethically. For instance, **Nestlé** uses blockchain to trace the supply chain of its coffee beans, guaranteeing sustainable farming practices.

- **Carbon Footprint Tracking:**

  Blockchain allows companies to measure carbon emissions at each stage of the supply chain. This data can be used to identify high-impact areas and implement targeted sustainability initiatives.

- **Waste Management Documentation:**

  Blockchain ensures accurate tracking of waste disposal and recycling processes, promoting accountability and reducing illegal dumping practices.

Future Trends in Sustainable Technologies

As the supply chain industry continues to evolve, future trends in sustainable technologies focus on achieving greater environmental efficiency, leveraging advanced digital tools, and adapting to global sustainability goals. These trends not only address environmental challenges but also enable businesses to remain competitive and resilient in a rapidly changing landscape.

1. Net-Zero Supply Chains

One of the most significant trends is the push toward net-zero supply chains, where businesses aim to eliminate or offset all carbon emissions generated by their operations.

- **Renewable Energy Integration:**

  Companies are investing heavily in renewable energy sources such as solar, wind, and geothermal power to run their logistics facilities and manufacturing plants. For instance, logistics hubs powered entirely by solar energy can significantly reduce operational emissions.

- **Carbon Offset Programs:**

  Many organizations are adopting carbon offset initiatives, such as afforestation and renewable energy credits, to balance out unavoidable emissions. Programs like reforestation projects and ocean cleanup initiatives align corporate goals with environmental restoration efforts.

- **Sustainable Shipping Practices:**

  The development of carbon-neutral shipping options, such as biofuel-powered vessels and electric cargo ships, is gaining momentum in global trade. Companies like **Maersk** have already launched vessels powered by biofuels as part of their decarbonization strategy.

2. Artificial Intelligence (AI)-Driven Sustainability

AI and machine learning are expected to play an even more prominent role in optimizing supply chain sustainability. These technologies will enhance decision-making by analyzing large datasets to identify inefficiencies, predict demand, and optimize resource allocation.

- **Predictive Analytics for Demand Management:**

  AI-powered models forecast demand with high accuracy, reducing overproduction and minimizing waste.

- **Energy Optimization Algorithms:**

  AI systems analyze energy consumption patterns across facilities and transportation networks, providing recommendations to reduce energy waste and optimize energy efficiency.

- **Dynamic Transportation Management:**

  AI improves transportation efficiency by dynamically adjusting routes based on real-time traffic, weather, and fuel consumption data, leading to reduced emissions.

3. Decentralized and Localized Supply Chains

The concept of decentralized and localized supply chains focuses on producing and distributing goods closer to the end customer. This trend reduces transportation distances, leading to lower fuel consumption and emissions.

- **Micro-Fulfillment Centers:**

  These smaller, localized warehouses are designed to serve urban areas efficiently, enabling faster delivery while reducing transportation requirements.

- **3D Printing for Local Production:**

  Additive manufacturing technologies such as 3D printing allow companies to produce customized goods on-demand at local facilities, minimizing waste and the need for global shipping.

4. Biodegradable and Smart Packaging

Packaging innovations are shifting toward materials that are both sustainable and intelligent.

- **Biodegradable Materials:**

  Packaging made from plant-based polymers, such as cornstarch or seaweed, decomposes naturally, reducing landfill waste.

- **Smart Packaging with IoT Integration:**

  IoT-enabled packaging includes sensors that monitor product freshness, temperature, and handling conditions, ensuring goods are delivered in optimal condition.

5. Green Hydrogen for Transportation and Operations

Green hydrogen, produced using renewable energy, is emerging as a clean and sustainable fuel source for powering heavy-duty vehicles, cargo ships, and even industrial operations.

- Hydrogen-powered trucks and buses are being developed for freight transportation, offering an alternative to fossil fuels.
- Warehouses and factories are exploring hydrogen fuel cells to power machinery, forklifts, and other equipment.

Strategies for Implementing Sustainable Technologies

Implementing sustainable technologies requires careful planning, collaboration, and a long-term vision. Below are detailed strategies to guide businesses in successfully integrating these technologies into their supply chains:

1. Conduct Comprehensive Sustainability Audits

- **Objective Assessment:**

  Businesses must begin by evaluating their current environmental impact across all areas of the supply chain, including energy consumption, emissions, and waste generation.

- **Identifying Key Focus Areas:**

  Audits help identify high-impact areas, such as energy-intensive warehouses or inefficient transportation networks, where sustainable technologies can be most effectively implemented.

- **Benchmarking and Goal Setting:**

  Organizations can use audit findings to set measurable sustainability goals, such as reducing carbon emissions by a specific percentage within a given timeframe.

2. Foster Collaboration Among Stakeholders

- **Engagement with Suppliers and Partners:**

  Companies must work closely with suppliers, logistics providers, and other partners to ensure alignment on sustainability goals and practices. For instance, requiring suppliers to use eco-friendly materials or adhere to ethical labor standards can enhance the entire supply chain's sustainability.

- **Joint Sustainability Initiatives:**

  Collaborative efforts, such as shared transportation networks or co-investments in renewable energy projects, can lead to cost savings and broader environmental benefits.

3. Invest in Employee Training and Development

- **Training Programs:**

  Employees must be educated on the importance of sustainability and trained to use new technologies effectively. This includes workshops on energy-efficient practices, reverse logistics management, and the operation of IoT-enabled systems.

- **Promoting a Culture of Sustainability:**

  Encouraging employees to suggest and implement green initiatives fosters a sense of ownership and accountability within the organization.

4. Leverage Financial Incentives and Government Support

- **Subsidies and Tax Benefits:**

  Many governments provide financial incentives for businesses adopting renewable energy, electric vehicles, and other green technologies. These incentives can help offset the initial costs of implementation.

- **Green Financing:**

  Collaborating with financial institutions to secure loans or grants for sustainability projects can ease the financial burden of adopting new technologies.

5. Develop a Phased Implementation Plan

- **Prioritizing High-Impact Areas:**

  Businesses should start by implementing sustainable technologies in areas with the most significant environmental impact, such as transitioning to EV fleets or retrofitting warehouses with renewable energy systems.

- **Scaling Gradually:**

  Once the initial phase proves successful, the organization can scale its sustainability initiatives across other areas of the supply chain.

6. Continuously Monitor and Optimize

- **Real-Time Monitoring Systems:**

  IoT devices and data analytics platforms enable businesses to track the performance of sustainable technologies, identify inefficiencies, and make data-driven improvements.

- **Periodic Reviews:**

  Regularly evaluating sustainability initiatives ensures alignment with changing regulations, technological advancements, and organizational goals.

Key Takeaways:

The integration of sustainable technologies into supply chain management is not just a response to global environmental challenges but also a pathway to greater resilience and profitability. Innovations like electric vehicles, IoT-enabled energy monitoring, blockchain for transparency, and renewable energy systems are transforming the industry, enabling businesses to meet sustainability goals without compromising operational efficiency.

As future trends such as net-zero supply chains, AI-driven sustainability, and decentralized supply chains gain traction, organizations must adapt by developing clear strategies for implementation. Businesses that proactively adopt sustainable technologies will benefit from reduced costs, enhanced brand reputation, and compliance with evolving regulations. By embracing this transformative shift, supply chains can play a pivotal role in creating a more sustainable, equitable, and prosperous global economy.

Case Study: Tata Steel's Integration of Circular Economy in Supply Chain Operations

Tata Steel, one of India's leading steel manufacturers, has set a benchmark in integrating sustainable technologies into its supply chain through the adoption of circular economy principles. The company's initiatives focus on reducing waste, conserving natural resources, and minimizing its carbon footprint across operations.

A key pillar of Tata Steel's strategy is the **reuse and recycling of industrial by-products**. The company has implemented advanced technologies to recover valuable materials from steelmaking waste. For instance, slag, a by-product of the steelmaking process, is repurposed for applications such as road construction and cement production. This approach not only reduces waste sent to landfills but also creates a secondary revenue stream while supporting other industries.

In addition, Tata Steel has invested in **energy-efficient production technologies** and renewable energy systems to reduce greenhouse gas emissions. By adopting high-efficiency equipment and process automation, the company has significantly lowered energy consumption in its manufacturing facilities. Furthermore, Tata Steel sources a portion of its power from renewable energy, aligning with its goal of achieving net-zero carbon emissions by 2045.

The company has also prioritized **responsible sourcing** by collaborating with suppliers to ensure raw materials are mined and transported sustainably. Tata Steel's logistics network incorporates fuel-efficient transportation methods and rail-based freight systems to reduce emissions associated with distribution.

These sustainability initiatives have produced tangible results. Tata Steel has reported a significant reduction in waste generation, increased recycling rates, and a notable decrease in its carbon footprint. The company's efforts demonstrate how integrating circular economy principles and sustainable technologies can enhance operational efficiency while promoting environmental stewardship.

This case study highlights Tata Steel's leadership in sustainability, serving as a model for industries seeking to build eco-friendly and resource-efficient supply chains.

References

Sarkis, J. (2019). *Handbook on the Sustainable Supply Chain*. Edward Elgar Publishing.

Srivastava, S. K. (2007). Green supply-chain management: A state-of-the-art literature review. *International Journal of Management Reviews, 9*(1), 53-80.

Seuring, S., & Müller, M. (2008). From a literature review to a conceptual framework for sustainable supply chain management. *Journal of Cleaner Production, 16*(15), 1699-1710.

Govindan, K., Soleimani, H., & Kannan, D. (2015). Reverse logistics and closed-loop supply chain: A comprehensive review. *Transportation Research Part E: Logistics and Transportation Review, 56*, 43-64.

Kumar, S., & Malegeant, P. (2006). Strategic alliance in a closed-loop supply chain, a case of manufacturer and eco-non-profit organization. *Technovation, 26*(10), 1127-1135.

Jabbour, C. J. C., Jabbour, A. B. L. S., Sarkis, J., & Govindan, K. (2013). Barriers to the adoption of green operational practices at Brazilian companies. *Business Strategy and the Environment, 22*(3), 182-201.

Christopher, M. (2016). *Logistics & Supply Chain Management*. Pearson.

Chopra, S., & Kalra, D. V. (2023). *Supply Chain Management: Strategy, Planning, and Operation*. Pearson.

Supply Chain Digital: https://supplychaindigital.com/

SCM World: https://www.scmworld.com/

World Economic Forum – Supply Chain and Transport: https://www.weforum.org/agenda/archive/supply-chain/

Harvard Business Review – Supply Chain Articles: https://hbr.org/topic/supply-chain

03

DIGITAL TRANSFORMATION OF SUPPLY CHAIN

About the Author - Miss Ananya Dhojia is a professional who has joined a renowned pharmaceutical industry to have extensive experience in the field of sourcing and supply chain management. She has a proven track record of success in procurement and supply chain management.

She is currently pursuing an MBA in logistics and supply chain management from Parul University and has a certificate of basic-level understanding of Blockchain and AI.

She has a deep understanding of supply chain and supplier auditing and a passion for sharing knowledge with aspiring professionals.

In logistics, many new technologies are used in developed countries, while in India adoption process is very slow. However, due to liberalization of the Indian economy, the competitive pressure is building up, and the only option to face the competition is to go in for technology-enabled operations. Supply chain management is crucial to the success of any company's operations and has a direct impact on its profits. More importantly, supply chain management approaches have a major impact on customer satisfaction, which outweighs decreased product costs. The improvement of SCM has been the fundamental source for enterprises to earn new profits and sustain their competitive edge. Every year, the SCM processes involving humans, nature, technology, and resources become more complex. For example, logistics processes must adapt to changes that occur as a result of continuous improvement, such as lean, total quality management, agility, and so on.

Robots, as an early branch of automation technology, were utilized for tedious and frequently dangerous jobs, including welding, painting, assembling, transportation, and explosive disposal. Robots have been used in increasingly complex tasks, including semi-autonomous control in agriculture, package sorting in logistics, and exploration and reconstruction mapping in rescue missions, thanks to advancements in technologies such as Robot Operating System (ROS), cloud robotics, and machine vision. However, the implementation of intelligent robots in such autonomous applications is still in its early stages. Many scholars and politicians have paid close attention to the use of technology in logistics and supply chains. In addition to increased human resource requirements and rising operational expenses, the trucking sector has encountered other significant obstacles. Over 75% of freight in the United States is transported by road, making the trucking industry the backbone of the country's transportation infrastructure. The trucking sector has had a labor shortage, with trucking businesses anticipating shortages of more than 175,000 truck drivers within a decade, given that more than 60% of truck drivers are 45 years or older.

However, autonomous vehicles (AVs) offer new chances to address these difficulties. By employing AVs in trucking, the demand for truck drivers may decrease while easing the shortage of truck drivers. Although the groundwork for AVs has been laid by many companies, such as Tesla, Google, and Uber, the application of AVs at the national level, specifically in truck freight transportation, is still limited.

Digital transformation is a broad word that refers to the conversion of a fully or partially analogue organization's activities to an upgraded digital environment. According to supply chain logistics activities, they rely heavily on IT, which is run in an innately digital context. Firms that distribute through Independent Warehouse Owners (IWOs) must adapt their basic activities to a new digital environment in order to effectively manage fleets and engines. Logistics and supply chain operations are being altered by technological developments and innovative technologies. The logistics and supply chain sector is experiencing a major transformation in procedures as it embraces emerging technologies such as the Internet of Things (IoT), blockchain, data analytics, artificial intelligence (AI), digital twins, and drones and robots.

Other technologies, such as validated data transfers and industrial models, have a substantial impact on these activities. Logistics, as a component of supply chain management, is concerned with the actual movement of raw materials and possessions from their point of origin in order to meet consumer demands. It also comprises managing and implementing transit strategies for incoming and outgoing goods, developing and coordinating the entry and flow of things and storage facilities, and many other tasks.

The logistics and supply chain industries have seen significant transformations as a result of fast technological innovation. This transformation has not only improved operations, but it has also altered the dynamics of global trade and client expectations. The integration of sophisticated technologies such as the Internet of Things

(IoT), artificial intelligence (AI), big data analytics, and automation has increased efficiency and responsiveness in this industry.

For starters, the deployment of IoT has streamlined supply chain management by allowing real-time tracking and monitoring of commodities. IoT devices, such as sensors and RFID tags, help to collect data across the supply chain. This data enables logistics managers to make more educated decisions about inventory levels, order processing, and delivery schedules, resulting in higher customer satisfaction. Efficient shipping tracking saves losses and theft while also allowing for proactive reactions to anticipated delays.

Furthermore, AI has altered the logistics industry by improving demand forecasting and inventory management. Machine learning algorithms can analyze previous data to accurately predict future demand patterns. This capability enables organizations to optimize their inventory levels while lowering holding costs. AI-driven analytics can help logistics organizations improve route planning and delivery timetables, leading to faster service and cheaper operational costs.

These technological advancements have an impact beyond operational efficiency; they also contribute to sustainability in the logistics and supply chain sectors. Companies can reduce their carbon footprint by optimizing routes and reducing fuel use through trustworthy data analytics. Furthermore, corporations are increasingly utilizing technology to create circular supply chains, in which waste is eliminated and resources are successfully reused, thus contributing to both economic and ecological sustainability.

However, integrating technology into logistics poses challenges. Concerns regarding data security and privacy have increased as consumers rely more on digital services. Logistics companies must invest in cybersecurity solutions to safeguard sensitive information while complying with data privacy regulations. Furthermore, ongoing workforce upskilling is essential to keep up with technological advances, ensuring that employees are well-prepared to manage and apply new technologies.

In the field of automatic identification, some of the more significant technologies are

Barcoding

Barcoding is a series of parallel lines of varying thicknesses with spaces between them. These bars are just codified information that can be read with a scanner. Historically, bar codes were introduced in supermarkets in the United States in 1952.The information written on a bar code includes the country code, manufacturer name, product specifications, date of manufacture, material content, and so on.These details are necessary at the user level for inventory management. Barcodes are utilized in a variety of industries, including retail, pharmaceuticals, consumer products, electronics, and cars.

Impact of Bar Code Technology on Operations of Logistics and Supply Chain Management –

i) Procurement operation – The parts and components brought from suppliers are assigned bar codes, which contain information on item name, batch number, date of manufacture, order number, serial number, etc. The information in the code helps in identifying and tracking the component. In the warehouse, when the goods enter through a conveyor, they are further scanned by the handheld scanner or scanner fixed alongside the conveyor. The information decoded by the scanner is immediately logged in the central computer, which helps with real-time updates of inventory records. ii) Processing – During the order processing, the bar code will help in keeping identification of items based on their date of entry into the warehouse or store. This will ease material storage, retrieval, and dispatch in a FIFO (First in First out) inventory management system. iii) Production operation – During the production process, the identification of in-process and finished items becomes easier due to barcoding. The various batches at different stages of production can be easily tracked. iv) Distribution operation – During distribution, a barcode helps in identifying and tracking the transit of finished goods to the customers.

Radio Frequency Identification (RFID)

RFID is an automatic identification and data capture (AIDC) technology. RFID originally debuted in tracking and access applications in 1980. RFID-based systems enable non-contact reading and are useful in manufacturing and other hazardous environments where bar codes would not survive. These are used as an alternative to barcodes to transmit inventory data to the reader via radio waves. RFID wirelessly transmits data between a tagged object and a reader.

a. RFID helps Indian exporters to global retailers like WAL-MART get better and more visibility into the movement of their goods within the supply chain and thus become more competitive.
b. Improve the ability of manufacturers to better manage the inventory levels.
c. Improve the complex distribution system for the defense operation.
d. Improve the complex tracking and distribution operations of the Indian Postal services.
e. Improve the tracking, logistics, and planning operations of Indian Railways and state public transport agencies.
f. Implement automatic toll collection on a vast network of highways.

RFTs

The reader is connected to the main computer. Radio Frequency Tags (RFTs) are silicon chips that store data in microcircuits. The RFTs are programmable and have erasable memory. Data is stored in coded form and transmitted to the reader via waves. The fundamental premise of tag is that the antenna emits radio signals. RFTs are quite beneficial for accompanying truck cargoes. The tag will include information on the consignor and consignee, inventory items, amount and value, the time the item traveled through a specific zone, and even the temperature. The reader receives the tag signal via its antenna, decodes it, and sends the data to the host computer system. RFTs can be attached to almost anything, including semitrailers, pallets, and containers. RFTs will avoid paperwork and can be helpful in quick clearance at octroi and customs posts. In the warehouse, the barcodes can be applied to the individual inventory items while RFTs can be applied to pallets, containers, etc. These will allow the staff to directly communicate with the warehouse computer. RFID has a significant impact on logistics and supply chains of many sectors.

Blockchain

Blockchain technology is a sophisticated database technique that enables transparent information sharing across a business network. A blockchain database holds data in blocks connected together in a chain. The data is chronologically consistent since deleting or modifying the chain requires network consensus. As a result, you can utilize blockchain technology to generate an unchangeable or immutable ledger for recording orders, payments, accounts, and other transactions. The system includes built-in measures to prevent unauthorized transaction entries and maintain consistency in the shared view of these transactions.

Case Studies: Blockchain in Supply Chain

- **Cargill:** Traceable turkeys
- **Carico coffee:** Traceable coffee
- **Starbucks:** Beans to cup
- **Walmart:** Blockchain traceable food supply chain
- **My story:** Traceable provenance of wine and products
- **Everledger:** Traceable diamonds from mine to consumer

IoT

The phrase IoT, or Internet of Things, refers to a network of linked objects and the technology that allows them to communicate with the cloud and with one another. We now have billions of internet-connected gadgets thanks to the development of low-cost computer chips and high-bandwidth telecommunications. This means that ordinary objects such as toothbrushes, vacuums, vehicles, and machinery can employ sensors to collect data and respond intelligently to user commands.

The Internet of Things connects everyday "things" to the internet. Since the 1990s, computer engineers have integrated sensors and processors into common things. The chips were large and thick, so progress was slow at first. Low-power computer chips known as RFID tags were initially employed to track pricey equipment. As computing devices shrank in size, these chips also became smaller, faster, and smarter over time.

The Internet of Things connects everyday "things" to the internet. Since the 1990s, computer engineers have integrated sensors and processors into common things. The chips were large and thick, so progress was slow at first. Low-power computer chips known as RFID tags were initially employed to track pricey equipment. As computing devices shrank in size, these chips also became smaller, faster, and smarter over time. The cost of embedding computational "ower" into small objects has now significantly decreased. For example, you can add connectivity with Alexa speech services capabilities to MCUs with less than 1MB of embedded RAM, such as light switches. A whole industry has built up around the idea of flooding our homes, companies, and offices with IoT gadgets. These smart items can autonomously send and receive data over the Internet. The Internet of Things refers to all of these "invisible computing devices" and the technology that connects them.

References

Baloch, Neelam, and Aamir Rashid. "Supply chain networks, complexity, and optimization in developing economies: A systematic literature review and meta-analysis: Supply chain networks and complexity: A meta-analysis." *South Asian Journal of Operations and Logistics*

Prajogo, D., & Sohal, A. (2013). Supply chain professionals: A study of competencies, use of technologies, and future challenges. *International Journal of Operations & Production Management*,

Sohal AS. Supply chain professionals: A study of competencies, use of technologies, and future challenges. International Journal of Operations and Production Management.

Bals, Lydia, et al. "Purchasing and supply management (PSM) competencies: Current and future requirements." *Journal of purchasing and supply management* 25.5 (2019): 100572.

Eslami, M. H., Jafari, H., Achtenhagen, L., Carlbäck, J., & Wong, A. (2024). Financial performance and supply chain dynamic capabilities: the Moderating Role of Industry 4.0 technologies. *International Journal of Production Research*, 62(22), 8092-8109.

04

INTERNET OF THINGS IN LOGISTICS

About the Author - Eng. Shiwesha Mwenda is a distinguished professional with a master's degree in logistics and supply chain management and a bachelor's degree in chemical engineering from Parul University, India. With extensive expertise and leadership experience, he serves as the General Manager of Operations, Logistics, and Transport at Giant Eagles Security Services and Car Hire Ltd, a prominent security firm in Lusaka, Zambia.

Beyond his professional achievements, Eng. Shikwesha is a published author of the acclaimed book "The Light of Darkness,' an inspirational and thought-provoking work offering unique wisdom and insights. His dedication to community leadership was evident during his tenure as president of the International Students' Union at Parul University (2021–2022), where he represented and advocated for a diverse student body.

As the CEO and founder of Mweshikwe Enterprise, Eng. Shikwesha continues to lead a fast-growing enterprise known for its commitment to quality and timely service delivery. A visionary leader with a passion for innovation and excellence, he remains a trailblazer in his field

Internet of Things in Logistics

The Internet of Things (IoT) is simply the "Eye That Sees Beyond Eyes" (By Eng. Shikwesha).

The IoT concept and term itself had been presented by Kevin Ashton the founder of the Auto-ID research group at the Massachusetts Institute of Technology in 1999. IoT integrates surrounding objects into a single network.

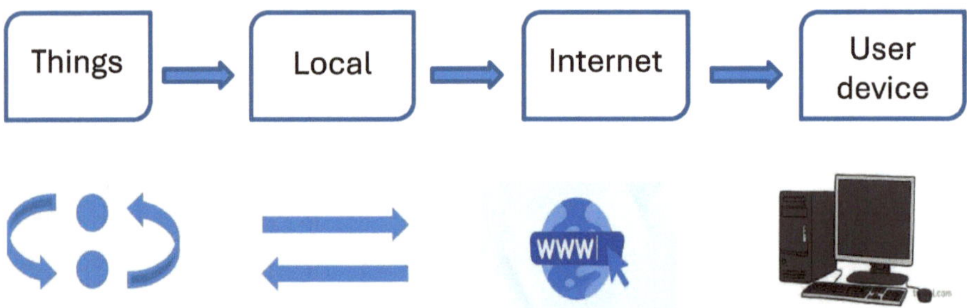

Figure 3.1

Internet of Things (IoTs): Refers to "a network of interconnected devices and sensors that enable machines to communicate with each other and centralized systems, facilitating real-time data collection, transfer, and analysis" (Atzori, Iera, & Morabito, 2010).

It's important to note that the logistics industry plays a critical role in meeting the needs of various sectors by facilitating the smooth movement of goods from production to consumption. Logistics is simply a key component in ensuring that materials (raw and finished goods) are available in the right quantity, at the right place, and at the right time, with zero or minimal errors and disruptions in the supply chain. Technically, logistics also ensures that goods reach their destination undamaged and at the lowest possible cost, making it a crucial element of successful supply chain management.

The benefits of IoTs in the logistics sector are that, by monitoring, tracking, and automating processes, IoT significantly enhances the efficiency, transparency, and accuracy of logistics operations, reducing human intervention and improving decision-making. In other words, IoT has emerged as a key enabler, revolutionizing logistics and supply chain management by enhancing visibility, operational efficiency, and well-informed decision-making processes.

The Internet of Things is also crucial in providing tangible solutions related to the long-standing logistical challenges such as poor infrastructure, high transportation costs, and limited last-mile connectivity.

The Role of the Internet of Things in Logistics

The logistics industry has long been characterized by fragmentation and inefficiency, but the rise of IoT is now offering a solution to these challenges. In today's era, where we have an increasingly competitive and digitalized landscape, it's essential to a greater extent to leverage IoT technology in ensuring unprecedented levels of transparency, resource allocation, real-time monitoring, warehouse automation, and optimize fleet management. IoT improves real-time visibility and monitoring of the flow of goods and information throughout the supply chain, from sourcing of raw materials to delivering finished goods to the customer. The proliferation of IoT in the logistics space is enabling firms to overcome traditional operational hurdles, such as poor connectivity networks and erratic delivery times.

Major Benefits of Internet of Things in Logistics

1. Increased Visibility:

IoT offers real-time tracking of products, inventory levels, and transportation. By embedding IoT technologies such as RFID, GPS tracking, and cloud-based data analytics, logistics companies are now achieving real-time visibility across the supply chain, from warehouse operations to last-mile delivery (Mckinsey & Company, 2021).

2. Optimization of Fleet Management:

Fleet Monitoring: IoT-enabled fleet management systems provide real-time vehicle tracking, monitor fuel usage, and send maintenance alerts to the management.

Predictive Maintenance: Sensors gather data on vehicle performance, enabling predictive maintenance, which is crucial in reducing downtime and optimizing fleet performance.

Cost Reduction: IoT helps reduce fuel consumption, repair costs, and operational inefficiencies. **Real-Time Alerts and Recovery Mode**: Switch to real-time tracking when theft or unauthorized movement is detected, improving the chances of asset recovery.

Route Optimization: IoT-based smart route planning helps minimize delivery times by analyzing traffic, weather, and road conditions.

Real-Time Tracking: IoT enables customers to track deliveries in real-time, and this plays a vital role in enhancing customer satisfaction levels.

Case Study (India): Blue Dart, a major Indian logistics company, reduced fuel consumption by 18% and vehicle downtime by 15% after deploying IoT-based fleet management solutions (Deloitte, 2020). This is just to show you how much relevance the IoT has brought to the logistics sector.

3. Warehouse Inventory Management:

RFID & IoT Sensors: IoT enhances warehouse management by automating inventory control using RFID and other sensors to track products in real-time.

Real-Time Stock Updates: IoT ensures that warehouse stock levels are constantly updated, reducing the likelihood of stockouts or overstocks, which would be costly to the firm if either of them occurred.

Temperature and Humidity Control: IoT sensors help maintain the correct environment for temperature-sensitive goods, such as chemicals, dairy products, and pharmaceuticals, reducing spoilage rate.

Efficiency Gains: With smart warehousing, Indian logistics companies have seen efficiency gains of up to 25% (Wamba et al., 2020).

4. Enhanced Operational Efficiency:

IoT's ability to streamline inventory management and predict maintenance needs is helping logistics firms achieve efficiency levels that were previously unattainable (KPMG, 2021).

IoT ensures faster decision-making through data analytics and real-time updates.

IoT-driven solutions are key to enhancing supply chain operations by reducing inefficiencies and human error by a substantial margin, which is essential for improving customer satisfaction levels.

5. Sustainable Environmental Benefits:

One of the most significant contributions of IoT in logistics is its ability to enhance environmental sustainability. By optimizing routes, monitoring fuel consumption, and minimizing waste through precise inventory management, IoT reduces the overall carbon footprint of logistics operations.

The McKinnon et al. (2015) suggest that smart freight transportation systems enabled by IoT can cut down on emissions and resource wastage, while the United Nations Development Program (UNDP, 2021) advocates for IoT as a critical component in achieving sustainable development goals, particularly in logistics where efficiency translates directly to environmental benefits.

6. Predictive Analytics:

The convergence of IoT with Artificial Intelligence (AI) is creating new opportunities for predictive analytics within the logistics sector. AI-powered IoT systems can anticipate supply chain disruptions, optimize inventory levels, and even predict equipment maintenance requirements, reducing downtime and ensuring smoother operations. Gartner (2021) highlights the role of predictive analytics in logistics, noting that the use of AI in conjunction with IoT systems enables logistics firms to make more informed, data-driven decisions, improving overall efficiency and cost-effectiveness.

Demerits of the Internet of Things

Like any other sector, the implementation of the IoT does have some disadvantages associated with it:

High Initial Costs: IoT systems require substantial upfront investments in sensors, software, and infrastructure.

Cybersecurity Risks: With the increased interconnectivity of devices, IoT systems are vulnerable to cyberattacks and data breaches.

Infrastructure Challenges: In India, inadequate infrastructure (e.g., unreliable internet, lack of tech integration) hinders full IoT implementation.

Case Example: Smaller logistics companies in India face budget constraints when trying to adopt IoT, which limits their competitiveness against larger players like Blue Dart (GSMA, 2021).

References

Atzori, L., Iera, A., & Morabito, G. (2010). The Internet of Things: A survey. Computer Networks, 54(15), 2787-2805.

Deloitte. (2020). IoT-based fleet management solutions reduce fuel consumption and improve efficiency for Blue Dart.

Gartner. (2021). Predictive analytics in logistics: Leveraging AI and IoT for operational efficiency.

GSMA. (2021). IoT adoption challenges among smaller logistics firms in India. KPMG. (2021).

Enhancing operational efficiency through IoT-driven logistics solutions .Mc Kinnon, A., et al. (2015).

Smart freight transportation systems enabled by IoT: Reducing emissions and resource wastage. Journal of Logistics and Sustainability Studies, 11(2), 123-138.

McKinsey & Company. (2021). The impact of IoT technologies on logistics and supply chain visibility.

United Nations Development Program (UNDP). (2021). Achieving sustainable development goals through IoT in logistics.

Wamba, S. F., et al. (2020). Efficiency gains through smart warehouse operations in the Indian logistics sector. International Journal of Logistics Management, 31(3), 456-472.

Ashton, K. (1999). The Internet of Things: Origin and conceptual framework. Massachusetts Institute of Technology, Auto-ID Research Group.

Blue Dart. (2020). Fleet management solutions: Success story of IoT implementation.

05

CLOUD COMPUTING IN LOGISTICS MANAGEMENT

About the Author - Mr. Manav Parashar is a logistics professional with a keen understanding of supply chain dynamics and software-driven solutions. Born in Madhya Pradesh and raised in Gujarat, Manav Parashar brings a practical and analytical approach to logistics, supported by academic excellence and real-world experience. With a B.Com in Marketing and an MBA in Logistics and Supply Chain Management, Manav Parashar combines their expertise with a passion for innovative problem-solving. They are skilled in integrating technology into logistics processes, ensuring efficiency and adaptability in ever-evolving markets. This chapter reflects Manav Parashar's insights into logistics software, offering practical strategies and perspectives that empower organizations to streamline their operations and enhance decision-making.

During my MBA, I focused on optimizing supply chain processes and efficiency. As part of my academic work, I published a research paper, which allowed me to explore advanced strategies for improving operational effectiveness in real-world business environments. This project further ignited my passion for studying and applying supply chain solutions, particularly in areas such as demand forecasting, inventory control, and process automation.

I am highly enthusiastic about staying updated on the latest trends, technologies, and innovations in supply chain management. I actively seek opportunities to deepen my knowledge and understanding of how supply chains can be made more efficient and responsive to the evolving demands of the global marketplace. My goal is to continue exploring new ways to optimize supply chains and contribute to enhancing business operations and sustainability.

Introduction

Brief Overview:

Cloud computing is a technology that enables the delivery of computing services—such as storage, processing power, networking, and software—over the internet. Instead of relying on physical infrastructure, cloud computing allows businesses and individuals to access these resources remotely, on demand, and typically on a pay-as-you-go basis. This model eliminates the need for large upfront investments in hardware and software, as well as the complexities of managing and maintaining on-site infrastructure.

The core advantages of cloud computing include scalability, flexibility, cost-effectiveness, and increased accessibility. With cloud solutions, organizations can scale their IT resources up or down based on demand, improve operational efficiency, and enable global access to data and applications. As businesses increasingly rely on

data-driven insights and real-time collaboration, cloud computing has become a critical enabler of innovation, driving digital transformation across industries.

Importance of Cloud Computing Specific to Logistics

Cloud computing has become a game-changer in the logistics industry, offering transformative benefits that enhance operational efficiency, decision-making, and customer satisfaction. Here's an expanded look at the importance of cloud computing for logistics:

- **Real-Time Data Access:**

Cloud computing enables logistics firms to access real-time data, such as shipment locations, inventory levels, and status updates. This capability allows businesses to track shipments in transit, providing timely and accurate information to customers. It also enhances decision-making, allowing logistics managers to identify issues like delays or bottlenecks and take corrective actions promptly. This leads to more efficient operations and better customer service.

- **Scalability:**

The cloud provides logistics companies with the flexibility to scale their IT resources up or down based on demand. During peak periods, such as holiday seasons, logistics companies can easily increase their storage capacity, processing power, and software functionalities. Conversely, during off-peak times, they can scale down to avoid unnecessary costs. This scalability ensures cost-effective operations and prevents companies from overinvesting in infrastructure that may not be needed year-round.

- **Centralized Data Management:**

Cloud-based systems centralize data across the entire supply chain, allowing stakeholders to access and share relevant information easily. Whether it's inventory data, transportation schedules, or customer feedback, centralized data management provides better visibility into operations. This improved integration leads to more accurate demand forecasting, optimized inventory levels, and smoother coordination between different stakeholders, reducing errors and delays.

- **Cost Efficiency:**

Cloud computing reduces the need for on-premise hardware, which can be expensive to purchase, maintain, and upgrade. By migrating to the cloud, logistics companies can eliminate the costs of server management, software licensing, and system updates. Instead, they pay for only the services and resources they use, allowing for more predictable costs and increased profitability. Additionally, the cloud's infrastructure is maintained and updated by the service provider, reducing the logistical burden on the company.

- **Enhanced Collaboration:**

The cloud allows seamless communication between logistics providers, shippers, suppliers, and customers. With a centralized platform, all stakeholders can access the same up-to-date information and collaborate in real time, making it easier to resolve issues quickly and coordinate deliveries. This improved collaboration leads to faster problem resolution and a higher level of service delivery, ultimately benefiting customers and strengthening business relationships.

- **Agility and Flexibility:**

Cloud computing empowers logistics companies to be agile and adapt quickly to changing market conditions. If a company faces sudden shifts in demand, regulatory changes, or supply chain disruptions, the cloud enables the flexibility to optimize operations in real time. Logistics firms can quickly adjust routes, allocate resources, or implement new strategies, which helps them stay competitive and meet customer expectations without delay.

- **Disaster Recovery and Backup:**

Cloud platforms typically offer built-in disaster recovery and backup solutions, ensuring business continuity even during disruptions. In case of natural disasters, cyberattacks, or technical failures, companies can quickly recover their data and resume operations without significant downtime. This level of preparedness is essential in the logistics sector, where delays or loss of data can lead to significant financial losses and damage to reputation.

- **Improved Customer Experience:**

Cloud computing enables logistics companies to offer enhanced services, such as real-time tracking of shipments and accurate estimated delivery times. Customers can access this information through online platforms, giving them more control over their orders and enhancing transparency. By keeping customers informed, logistics companies can boost satisfaction and loyalty, which is critical for long-term success.

- **Data Analytics:**

The cloud provides powerful tools for data analytics, enabling logistics companies to optimize routes, improve inventory management, and enhance overall operational efficiency. By analyzing historical data, companies can identify patterns, predict demand fluctuations, and fine-tune their processes. This not only improves efficiency but also reduces costs, such as fuel and labor, while optimizing the use of assets like trucks and warehouses.

Why Cloud Computing Over Traditional Approach?

Aspect	Cloud Computing	Traditional Approach
Deployment and Infrastructure	Hosted on remote servers managed by third-party providers; accessed over the internet	Hosted on-premises; businesses manage their own physical infrastructure (servers, storage)
Cost Structure	Pay-as-you-go model; pay only for what is used, reducing upfront costs	High upfront capital expenditure for hardware and software, plus ongoing maintenance costs
Scalability	Easily scalable, resources can be increased or decreased based on demand	Limited by physical hardware; scaling requires additional investments in infrastructure
Management and Maintenance	Provider handles management, maintenance, updates, and security	Business is responsible for managing hardware, software, security, and troubleshooting

Aspect	Cloud Computing	Traditional Approach
Access and Flexibility	Accessible from anywhere with an internet connection, supporting remote work and global collaboration	Typically restricted to on-premises infrastructure; remote access requires complex setup (e.g., VPN)
Security and Backup	Advanced security features (encryption, multi-factor authentication, automatic backups) managed by the cloud provider	Security and backup depend on the business's internal systems and practices
Innovation and Agility	Quick adoption of new technologies (AI, big data, etc.) and tools with minimal setup	Innovation is slower, often requiring significant investment in new hardware or software
Global Reach	Global infrastructure with data stored closer to users for better performance	Expanding globally requires building physical infrastructure in new regions, which is resource-intensive

Types of Cloud Computing Solutions:

Cloud computing solutions are typically categorized into three primary service models, each offering different levels of control, flexibility, and management responsibilities. These models include Infrastructure as a Service (IaaS), Platform as a Service (PaaS), and Software as a Service (SaaS). Additionally, cloud solutions can be deployed in different environments, such as public cloud, private cloud, and hybrid cloud. Below is a detailed overview of each type of cloud computing solution:

1. Infrastructure as a Service (IaaS)

Overview: IaaS provides the most fundamental layer of cloud services, offering virtualized computing resources over the internet. With IaaS, users can rent IT infrastructure, including virtual machines, storage, networking, and servers, without having to purchase or maintain physical hardware.

Key Components:

- Compute: Virtual machines that can be scaled based on demand.
- Storage: Scalable storage options like object storage and block storage.
- Networking: Virtual networks, load balancers, and VPNs.
- Operating Systems: Choose from a range of operating systems to install on virtual machines.

Examples of IaaS Providers:

- Amazon Web Services (AWS): Offers a wide range of computing resources such as EC2 instances for computing, S3 for storage, and VPC for networking.
- Microsoft Azure: Provides services like Virtual Machines, Azure Storage, and Azure Networking.
- Google Cloud Platform (GCP): Offers Compute Engine for virtual machines, Cloud Storage, and Cloud Networking.

Use Cases:

- Startups and SMEs: Ideal for businesses that need to scale infrastructure quickly but want to avoid upfront costs of physical hardware.
- Testing and Development: Provides a flexible platform to test new applications or services without investing in on-premise hardware.

- Disaster Recovery: Businesses can set up backup and recovery systems without maintaining separate physical infrastructure.

2. Platform as a Service (PaaS)

Overview: PaaS provides a platform that allows developers to build, deploy, and manage applications without worrying about underlying infrastructure. It abstracts the infrastructure layer and focuses more on software development, offering tools, libraries, and services to streamline app creation.

Key Components:

- Development Tools: Integrated development environments (IDEs), version control systems, and debugging tools.
- Middleware: Software that connects the application to databases and other services (e.g., databases, messaging queues).
- Database Services: Managed databases (SQL and NoSQL), caching solutions, and data storage.
- Application Hosting: The platform allows you to deploy applications directly.

Examples of PaaS Providers:

- Heroku: A popular platform for developers to deploy and manage applications, offering easy integration with various programming languages and databases.
- Google App Engine: Provides a fully managed platform to build and deploy applications without managing the underlying infrastructure.
- Microsoft Azure App Services: A platform offering services like web apps, mobile apps, and API apps for development.

Use Cases:

- Application Development: Ideal for businesses looking to develop custom applications without managing hardware or software stacks.

- Collaboration and Team Development: Developers can collaborate easily using the shared PaaS tools and environment, speeding up the development lifecycle.
- Mobile and Web Apps: PaaS is commonly used for building scalable, high-performance web and mobile applications.

3. Software as a Service (SaaS)

Overview: SaaS provides fully managed software applications hosted on the cloud. Users access these applications through a web browser, eliminating the need for installation, maintenance, or upgrades. SaaS solutions are often subscription-based and are ready-to-use applications.

Key Components:

- Applications: Fully functional software tools for specific tasks (e.g., CRM, email, accounting).
- Access via Browser: SaaS applications are accessible from any device with an internet connection.
- Automatic Updates: Software is updated automatically by the provider without any user intervention.

Examples of SaaS Providers:

- Salesforce: A leading CRM (Customer Relationship Management) platform that offers a suite of tools for sales, marketing, and customer service.
- Google Workspace (formerly G Suite): Includes productivity tools like Gmail, Google Docs, and Google Drive.
- Microsoft 365: A subscription service offering cloud-based Office applications such as Word, Excel, and PowerPoint, along with collaboration tools like Teams.

Use Cases:

- Business Software: Companies use SaaS solutions for enterprise resource planning (ERP), customer relationship management (CRM), project management, and more.

- Collaboration: SaaS tools like Google Workspace and Microsoft 365 enable remote teams to collaborate in real time.
- Email & Communication: Services like Gmail, Microsoft Outlook, and Slack are widely used for communication and file sharing.

4. Cloud Deployment Models

In addition to the service models (IaaS, PaaS, and SaaS), cloud computing solutions can be deployed in different environments based on security, compliance, and business needs. These environments are:

Public Cloud

- Overview: Public cloud services are owned and operated by third-party cloud providers, and resources like servers and storage are shared among multiple organizations. These clouds are accessible via the internet, and companies do not need to maintain the infrastructure themselves.
- **Examples:** AWS, Microsoft Azure, Google Cloud.
- **Use Cases:**
  - Startups and SMBs: For businesses looking to minimize capital expenditure and scale as needed.
  - Flexible, cost-effective resources: Great for businesses with variable workloads and the need for high scalability.

Private Cloud

- Overview: A private cloud is a cloud infrastructure dedicated to a single organization. It can be hosted either on-premises or by a third-party provider. It offers more control over data, security, and compliance, making it ideal for businesses with stringent data requirements.
- **Examples**: VMware vSphere, OpenStack.

- **Use Cases:**

    o   Highly regulated industries: Ideal for healthcare, finance, and government organizations that require enhanced security and compliance.
    o   Customization: Suitable for businesses that need more control over their infrastructure and data security.

Hybrid Cloud

- Overview: A hybrid cloud combines both public and private cloud solutions, allowing data and applications to be shared between them. This provides businesses with greater flexibility to run workloads in the most appropriate environment, optimizing costs, security, and performance.
- **Examples:** Microsoft Azure Hybrid, AWS Outposts.
- **Use Cases:**

    o   Workload Optimization: Companies can run critical workloads in a private cloud while leveraging the scalability of public cloud for less sensitive tasks.
    o   Disaster Recovery: Hybrid clouds are commonly used for backup and recovery solutions, ensuring business continuity by using the public cloud as a secondary infrastructure.

Conclusion

Cloud computing solutions offer businesses a range of options to address different needs:

- IaaS provides scalable and flexible infrastructure.
- PaaS focuses on delivering platforms for developers to build and deploy applications.
- SaaS offers complete software applications for end users, eliminating the need for maintenance and upgrades.

These solutions can be deployed in different environments, such as public, private, or hybrid clouds, depending on the organization's requirements for security, control, and cost-effectiveness. The right combination of these cloud services enables businesses to be more agile, reduce costs, and scale efficiently.

Example

A Logistics Business Transformed: From Tradition to Cloud Computing

Introduction

Rajesh Patel, the owner of **Patel Logistics**, a reputable logistics company operating across India, had built his business over several years. However, despite the company's long-standing success, Rajesh was facing increasing challenges in maintaining operational efficiency with outdated IT systems and fragmented processes. His company was still relying on traditional methods, which were no longer capable of supporting the growing demand and complexity of modern logistics operations. In order to stay competitive in an industry rapidly adopting new technologies, Rajesh realized that he needed to embrace cloud computing to streamline operations and improve his services.

Struggles of Traditional Logistics

Patel Logistics had built a solid foundation, but the traditional methods it relied on began to expose inefficiencies and operational bottlenecks that hampered growth. Some key issues faced by the company included:

- **Fragmented Operations:** Data was scattered across multiple systems and locations, leading to a lack of synchronization between departments. The inability to access data in real-time made it difficult to track shipments, manage inventories efficiently, or respond quickly to customer inquiries. Manual processes, such as paper-based reports and spreadsheets, further slowed operations.

- **Inefficient Fleet Management:** The fleet of trucks used by Patel Logistics relied on basic GPS systems and manual route planning. This often led to delays, underutilization of resources, and high fuel costs, as routes were not optimized for efficiency. Drivers were not always updated on the best routes in real time, causing unnecessary detours and increased delivery times.

- **Limited Customer Visibility:** Customers were left in the dark when it came to tracking their shipments. The absence of a centralized tracking system meant that customers had to rely on periodic updates, resulting in frustration and a negative customer experience.

- **High Maintenance Costs:** Maintaining and upgrading on-premise IT infrastructure was becoming increasingly expensive. The need for constant software updates, hardware repairs, and troubleshooting added to the company's operational costs.

Recognizing these challenges, Rajesh knew that modernizing his company's operations through technology was crucial to maintaining a competitive edge in the logistics industry.

The Decision to Move to Cloud Computing

Rajesh's turning point came at an industry event where he met Anita, a business owner who had successfully transitioned her logistics company to the cloud. Anita shared her experience and the tangible benefits she had seen from adopting cloud computing. Some of the key advantages she highlighted included:

- **Centralized Data Management:** With cloud computing, all data was accessible from a single, unified platform, ensuring that every department could access real-time information. This centralization removed the need for duplicate systems, improved data consistency, and facilitated faster decision-making.

- **Scalability:** Cloud-based solutions offered flexibility to scale IT resources up or down according to demand. This was particularly important during peak seasons, like the holiday rush, when logistics companies typically experience a surge in business. Instead of investing in additional physical infrastructure, Rajesh could now expand his operations quickly and cost-effectively by simply increasing his cloud resources.

- **Improved Route Optimization:** Cloud platforms enabled real-time data integration, allowing for better fleet management. With real-time traffic information, weather updates, and route optimization algorithms, Rajesh's fleet could take the most efficient routes, reducing delays, saving fuel, and ensuring timely deliveries.

- **Customer Transparency:** With cloud-based tracking systems, customers could track their shipments in real time, improving transparency and customer satisfaction. This shift from manual updates to automated tracking provided customers with accurate information, reducing their uncertainty and enhancing their experience.

After hearing these benefits, Rajesh decided it was time to modernize Patel Logistics and move his operations to the cloud.

Implementing the Cloud Solution

To begin the transformation, Rajesh partnered with a cloud service provider that specialized in logistics solutions. The transition to the cloud involved several important steps:

- **Employee Training:** Since the staff at Patel Logistics was accustomed to traditional systems, Rajesh ensured that employees received adequate training on the new cloud-based systems. The user-friendly nature of cloud software helped employees quickly adapt to the changes, reducing resistance and minimizing the learning curve.

- **Data Migration:** A significant part of the transition was migrating existing data from legacy systems to the cloud. Rajesh worked with the cloud provider to centralize all of Patel Logistics' data, ensuring that every department—from warehouse management to fleet operations—had real-time access to the information they needed.

- **Cloud-Based Fleet Management:** One of the key components of the cloud solution was the introduction of a Transportation Management System (TMS), which was hosted on the cloud. This system allowed Rajesh to optimize delivery routes based on real-time data and monitor the performance of his fleet. It also provided automated alerts for any delays or issues, allowing the team to respond proactively.

The Benefits of Cloud Computing

After the transition, Patel Logistics began to experience a host of improvements that had a positive impact on both operational efficiency and customer satisfaction:

- **Increased Efficiency:** With cloud computing, the company could now operate more efficiently. Employees from different departments could access the same data simultaneously, improving collaboration. Shipment tracking was instant, and information flowed seamlessly between departments, ensuring smooth operations.

- **Cost Savings:** By eliminating the need for costly hardware, software upgrades, and IT infrastructure maintenance, Patel Logistics saw a significant reduction in operating costs. Additionally, cloud-based solutions required less in-house IT support, further lowering overhead costs.

- **Better Fleet Management:** With cloud-powered route optimization, Patel Logistics was able to reduce fuel consumption, minimize delays, and improve overall fleet efficiency. The TMS provided insights into vehicle performance, helping Rajesh manage his fleet more effectively and extend the life of the vehicles.
- **Enhanced Customer Experience:** Real-time tracking and more accurate delivery timelines improved the customer experience. Customers could easily track their shipments via a user-friendly portal, resulting in higher levels of satisfaction and loyalty.
- **Scalable Growth:** The cloud provided Patel Logistics with the flexibility to scale up operations without the need for substantial investments in physical infrastructure. As the company grew, it could easily add new users, increase storage, or incorporate additional features without significant upfront costs.

Looking Forward

Six months after the transition, Rajesh reflected on the profound impact cloud computing had on his business:

- **Data-Driven Decisions:** With access to real-time data and advanced analytics, Patel Logistics was now able to make more informed decisions. Operations were continually optimized based on up-to-date information, and Rajesh could foresee trends and proactively address potential issues.
- **Higher Profit Margins:** The reduction in operational costs, combined with more efficient fleet management and faster deliveries, resulted in higher profit margins. The increased efficiency also allowed Patel Logistics to serve more customers without hiring additional staff, contributing to overall profitability.
- **Future Growth:** The flexibility of the cloud enabled Patel Logistics to explore new opportunities, such as expanding into new regions or introducing additional services. Rajesh now felt confident that his company could scale up rapidly to meet future demands.

Conclusion

Rajesh Patel's decision to shift to cloud computing transformed Patel Logistics from a traditional, fragmented operation to a modern, efficient, and scalable business. By embracing cloud technology, Rajesh overcame inefficiencies, reduced operational costs, and significantly enhanced the customer experience. The flexibility and scalability provided by the cloud have positioned Patel Logistics for sustained growth and success, enabling the company to remain competitive in the evolving logistics market. This story highlights that even long-established businesses can thrive through technological innovation, ensuring they remain relevant in an increasingly digital world.

Cloud computing has proven to be a game-changer across various industries, and logistics is no exception. By offering flexibility, scalability, and cost efficiency, cloud technology empowers businesses to streamline operations, improve customer satisfaction, and enhance overall performance. Its ability to centralize data, optimize resources, and enable real-time decision-making makes it an invaluable tool for modernizing logistics companies. As we've seen in the example in this chapter, cloud computing not only addresses the inefficiencies of traditional systems but also provides the agility required to stay competitive in a rapidly evolving market. Embracing cloud technology is no longer a choice but a necessity for logistics companies aiming for growth and success in the digital era.

References

1. Cao, Q., Schniederjans, D. G., & Schniederjans, M. J. (2017). "Establishing the use of cloud computing in supply chain management." Operations Management Research, 10*(1-2), 47-63.

   – This paper discusses the role of cloud computing in enhancing supply chain efficiency, with a focus on cost reduction and agility.
   – [Link](https://link.springer.com/article/10.1007/s12063-017-0123-6)

2. Tiwari, S., Jain, A., & Pandey, A. (2013). "Analysis of Supply Chain Management in Cloud Computing." *International Journal of Innovative Technology and Exploring Engineering (IJITEE), 3*(5), 113-117.

   – Examines the integration of cloud computing in SCM and its impact on transparency and efficiency.
   – [Link](https://www.ijitee.org/wp-content/uploads/papers/v3i5/E1286103513.pdf)

3. Gayialis, S. P., Kechagias, E. P., Deligianni, A., Konstantakopoulos, G. D., & Papadopoulos, G. A. (2022)."Implementation Technologies of an Advanced Cloud-based System for Distribution Operations." *arXiv preprint arXiv:2202.13954.*

   – This study highlights the advantages of cloud-based solutions for warehouse and distribution operations.
   – [Link](https://arxiv.org/abs/2202.13954)

4. Benotmane, A., Benabdelhafid, A., & Bouksour, O. (2018). "Perception towards Cloud Computing in Logistics Company." *International Journal of Academic Research in Business and Social Sciences, 8*(12), 1-15.

   – Provides insights into the adoption challenges and benefits of cloud computing in logistics.
   – [Link](https://hrmars.com/papers_submitted/14166/perception-towards-cloud-computing-in-logistics-company.pdf)

5. Trappey, A. J. C., Trappey, C. V., Govindarajan, U. H., Sun, J. J., & Chuang, A. C. (2017). "A review of technology standards and patent portfolios for enabling cyber-physical systems in advanced manufacturing." *IEEE Access, 4*, 7356-7382.

   – Discusses cloud computing as a part of Industry 4.0 and its implications in logistics and supply chains.

6. Sornprom, N. (2024)."Role of Cloud Computing & Artificial Intelligence in the Logistics & Supply Chain Industry." *Transactions on Engineering and Computing Sciences, 12*(6), 1-13.

   – Explores AI-driven cloud solutions for improving supply chain decision-making.

7. Hung, Y.-H. (2019)."Cloud Computing: A Study of Logistics as a Service (LaaS)." *International Journal of Future Computer and Communication, 8*(2), 54-60.

   – Introduces the concept of Logistics-as-a-Service (LaaS) and its implementation using cloud computing.
   – [Link](https://www.ijfcc.org/vol8/546-M037.pdf)

06

AUGMENTED AND VIRTUAL REALITY IN LOGISTICS TRAINING

About the Author - Mr. Krish Dadwani is currently pursuing an MBA with a specialization in Logistics and Supply Chain Management (LSCM) at the Parul Institute of Research and Management, Vadodara. He holds a Bachelor of Engineering (BE) degree in Computer Science. His strong foundation in technology, paired with his focus on supply chain management, enables him to approach logistics challenges with an analytical and data-driven mindset.

Krish gained practical industry experience as a sales executive at a leading freight forwarding company, Oceania World, where he worked for one year. During this time, he was responsible for managing client relationships, understanding customer logistics needs, and driving business growth through effective sales strategies. His role gave him valuable insights into the intricacies of the freight forwarding industry and the importance of efficient logistics operations in global supply chains.

1. Introduction

As the logistics industry continues to evolve at a fast pace, the demand for effective training solutions has never been greater. Traditional approaches like on-the-job training, classroom sessions, and instructional videos often fall short in terms of engagement, cost-effectiveness, and overall efficiency. To overcome these limitations, Augmented Reality (AR) and Virtual Reality (VR) are transforming the way logistics companies train their employees.

Augmented Reality (AR) enriches the real world by superimposing digital information onto the physical environment, enabling workers to make better, data-driven decisions in real-time. For instance, AR can display real time, step-by-step guidance, helping employees streamline their tasks, reduce mistakes, and improve overall efficiency. In contrast, virtual reality (VR) offers an entirely immersive, simulated experience, allowing workers to practice complex or hazardous tasks in a virtual setting, free from the dangers of real-world execution. VR training scenarios can simulate critical logistics operations such as operating forklifts, managing hazardous materials, or responding to emergencies, giving trainees the opportunity to rehearse high-risk activities without exposing themselves to potential harm or costly errors.

AR and VR technologies deliver interactive, immersive training experiences that can greatly improve learning effectiveness. By offering safe, affordable, and scalable solutions, these technologies help workers develop their skills, boost operational efficiency, and reduce the likelihood of errors. This chapter delves into how AR and VR are applied in logistics training, highlighting their advantages, challenges, and future possibilities.

1.1 Applications of AR and VR in Logistics Training

A. Warehouse and Inventory Management

In warehouse training, AR can be leveraged to improve real-time operations. For instance, smart glasses with AR capabilities can guide workers by displaying instructions for locating, picking, and packing items efficiently. These glasses can also offer visual directions to the correct aisle or shelf, minimizing errors and boosting productivity.

For VR, entire warehouse environments can be recreated virtually, allowing employees to practice navigating the space, operating equipment like forklifts, and managing inventory without the need for physical presence. This approach not only cuts costs but also provides a safe, controlled setting for training.

Suggested Diagram: A warehouse environment with AR/VR integration (showing a worker with smart glasses or VR headset).

B. Forklift and Equipment Training

Operating heavy machinery, such as forklifts, can pose significant risks if operators are not properly trained. VR simulations create a lifelike environment where employees can practice operating forklifts safely and effectively. These virtual scenarios can simulate challenging conditions, such as navigating tight spaces or dealing with adverse weather, which workers may face in real-world situations.

In contrast, AR can support equipment maintenance and troubleshooting. Using AR glasses or mobile applications, workers can receive detailed, step-by-step instructions for performing routine inspections, fixing issues, and ensuring machinery is running smoothly. This real-time guidance enhances productivity and minimizes downtime.

Suggested Diagram: Forklift and Equipment Training with AR/ VR integration (showing a worker with smart glasses or VR headset).

C. Supply Chain and Logistics Process Training

For more intricate logistics tasks, like supply chain management, both AR and VR are utilized to simulate entire supply chains, from production to delivery. VR enables employees to virtually explore different stages of the supply chain, helping them understand how various elements— such as transportation routes, customs regulations, and warehouse layouts—impact the overall process. This type of training is crucial for workers who need to grasp logistics operations on a global scale.

On the other hand, AR allows logistics professionals to access real-time information about the supply chain. For example, AR systems can provide warehouse workers with the latest updates on order statuses, shipments, and inventory levels by simply scanning a bar code or package, enabling faster and more accurate decision-making.

D. Customization and Professionalization of Training Programs

Customized Training Scenarios: AR and VR enable businesses to design training programs tailored to specific roles, work environments, or operational requirements. For example, a warehouse employee can participate in a personalized simulation that mirrors the company's actual warehouse layout and processes. This level of customization ensures that training is highly relevant to the employee's daily tasks, improving both knowledge retention and training outcomes.

Dynamic Learning Paths: Sophisticated AR and VR systems can adjust training based on the learner's progress. If an employee is having difficulty with a particular skill or task, the system can offer additional practice exercises or alter the training scenarios to target the areas where improvement is needed.

E. Global Training with Remote Access Remote Training and Team

Collaboration: One of the key benefits of AR and VR is their ability to facilitate training across different geographic locations. Employees can participate in training sessions from any location, and VR platforms can enable virtual collaboration, allowing workers to interact and learn together despite being physically distant. This is especially valuable for multinational companies with offices in various regions, as it helps maintain uniform training standards across the organization.

Cost Savings for Global Teams: Rather than incurring expenses for employee travel to central training centers, companies can use AR and VR for remote training, significantly reducing travel and lodging costs. Virtual training environments or workshops can replicate real-world logistics scenarios, providing employees with the same hands-on experience as if they were on-site.

1.2 Challenges and Limitations of AR and VR in Logistics Training

A. High Initial Investment

Although AR and VR can offer cost savings over time, the initial investment can be quite high. Companies must purchase the necessary hardware (such as headsets and smart glasses) and software (including simulation programs and AR applications). For smaller businesses, this upfront cost can present a significant obstacle to adopting these technologies.

B. Technical Limitations

For VR, the effectiveness of simulations relies on the technology's level of fidelity and precision. Poor-quality VR experiences can result in unrealistic training scenarios, reducing the overall impact of the training. Similarly, AR depends on advanced software and hardware to operate effectively, and it may experience technical issues or connectivity problems that can disrupt the training process.

C. Resistance to Change

Certain employees and managers might be hesitant to adopt new technologies, particularly older workers who may be less accustomed to AR and VR tools. To facilitate the transition, companies must provide thorough onboarding and ongoing support to help employees adjust to these innovative training methods.

1.3 Benefits of AR and VR in Logistics Training

1. Immersive Learning Experience

One of the key benefits of AR and VR is the immersive, interactive nature of the learning they provide. Unlike traditional methods, such as reading from textbooks or watching passive video tutorials, AR and VR engage learners in hands-on practice. This experiential learning allows workers to actively participate and learn by doing, creating opportunities to make mistakes in a controlled environment, which speeds up the overall learning process.

2. Cost-Efficient and Calculable Solutions

Logistics training often involves high costs related to physical equipment, real-world simulations, and significant time investments. AR and VR offer cost-effective alternatives by providing calculable solutions. For instance, VR simulations enable multiple employees to train simultaneously without requiring physical equipment. Similarly, AR can minimize the need for physical inventory or training materials, resulting in reduced operational costs.

3. Minimizing Human Error and Enhancing Safety

Human errors in logistics can lead to costly mistakes, accidents, and delays. AR and VR training programs help mitigate these risks by offering precise, step-by-step instructions for complex tasks. VR simulations also allow workers to practice high-risk activities, such as operating machinery, in a safe, virtual setting, thereby lowering the risk of accidents during hands-on training.

4. Instant Performance Feedback

AR systems are capable of providing real-time feedback during training sessions. For example, a worker may receive immediate alerts if they deviate from the correct procedures or make an error. This instant feedback reinforces correct actions, helping employees continuously improve their skills and enhance their performance.

Conclusion

Augmented and virtual reality are revolutionizing logistics training by providing immersive, affordable, and calculable learning opportunities. These technologies empower workers to gain hands-on experience while enhancing safety, efficiency, and precision in various logistics tasks.

Although challenges such as cost and technical integration still exist, the advantages of AR and VR are clear. As technology continues to evolve, AR and VR are expected to become essential tools for logistics companies aiming to maintain a competitive edge in an ever-changing industry.

References

Dörner, D., Grimm, P., & Schweiger, G. (2019). *Virtual Reality in Logistics and Training.* **Springer.** *An in-depth look at how virtual reality can be applied to logistics operations and training programs.*

Schmalstieg, D., & Hollerer, T. (2016). *Augmented Reality: Principles and Practice.* **Addison-Wesley Professional.** *A comprehensive guide that explains the fundamental concepts and practical applications of augmented reality.*

Jerald, J. (2015). *The VR Book: Human-Centered Design for Virtual Reality.* **Morgan & Claypool.** *Focuses on designing VR experiences with an emphasis on human factors, making it a valuable resource for VR training design.*

Kipper, G., & Rampolla, J. (2012). *Augmented Reality: An Emerging Technologies Guide to AR.* **Elsevier.** *Provides an overview of AR technology, its development, and its diverse applications, including training contexts.*

Aldrich, C. (2009). *Learning in 3D: Adding a New Dimension to Enterprise Learning and Collaboration.* **John Wiley & Sons.** *Explores immersive learning environments and the role of 3D simulations, offering insights relevant to VR-based training solutions.*

Papagiannis, H. (2017). *Augmented Human: How Technology Is Shaping the New Reality.* **Wiley.** *Examines the impact of emerging technologies like AR on our daily lives and professional environments, including training and logistics.*

SUSTAINABLE TECHNOLOGIES IN GREEN LOGISTICS

About the Author - Miss Shweta Ulhas Amilkanthwar is currently pursuing an MBA in LSCM (Logistics and Supply Chain Management) from Parul University, Vadodara. Holds a Bachelor of Commerce degree from SSR College of Arts, Commerce, and Science, Silvassa.

Always been eager to expand knowledge and explore new opportunities, which led to completing an internship at **Welspun Living Ltd, Vapi**.

Which benefited to gain practical insights into the shipping and logistics department. Who loves to learn from everyone she meets and apply that knowledge to enhance her personal and professional growth. With a passion for continuous learning and improvement, she is determined to make a meaningful contribution to the field of supply chain management.

Introduction to Green Logistics

Green logistics refers to the practices, strategies, and technologies that aim to reduce the environmental footprint of logistics and supply chain operations. It encompasses a variety of solutions designed to optimize resource use, minimize waste, and lower greenhouse gas emissions associated with the transportation, storage, and distribution of goods. In an era where climate change, resource depletion, and sustainability have become pressing global issues, green logistics represents an essential shift toward more eco-friendly, cost-efficient, and socially responsible logistics practices.

The logistics industry is a major contributor to environmental degradation, particularly due to its reliance on fossil fuels, high energy consumption, and waste generation. As the demand for goods continues to grow, there is an increasing need for sustainable technologies that can address these challenges. Green logistics involves a wide spectrum of innovations, from adopting energy-efficient transport modes to the implementation of smart technologies that enhance supply chain visibility and optimization.

Green logistics also emphasizes circular economy practices, which focus on recycling, repurposing, and reusing materials and products. By adopting packaging innovations and reverse logistics, businesses can reduce waste and increase the lifespan of products and materials, further contributing to sustainability.

Introduction to Sustainable Technologies in Green Logistics

Sustainable technologies in green logistics refer to the innovative solutions and systems designed to reduce the environmental impact of logistics and supply chain operations. These technologies focus on optimizing energy use, minimizing waste, and lowering carbon emissions throughout the transportation, warehousing, and distribution processes. As the world faces growing concerns over climate change and resource depletion, sustainable technologies offer a way to balance economic growth with environmental responsibility.

Key advancements include the use of alternative fuel vehicles, such as electric and hydrogen-powered trucks; energy-efficient warehouses; and smart technologies like artificial intelligence (AI) and the Internet of Things (IoT) to optimize routes and inventory management.

Additionally, renewable energy sources like solar and wind power are increasingly being integrated into logistics operations, further reducing reliance on fossil fuels. These technologies not only help companies meet environmental regulations but also contribute to cost savings, improved operational efficiency, and a more sustainable global supply chain.

This chapter explores the key sustainable technologies currently reshaping green logistics, examining their implementation, benefits, and challenges. We will explore how these innovations are contributing to a more sustainable logistics system and the ways in which businesses can leverage these technologies for long-term environmental and operational gains.

1. Electric and Hybrid Vehicles: Reducing Emissions in Transportation

Transportation is one of the largest sources of greenhouse gas emissions in logistics. Traditionally reliant on fossil fuels, the sector has seen a significant shift towards electric and hybrid vehicles (EVs and HEVs) as part of green logistics strategies.

Electric Vehicles (EVs):

Electric vehicles are increasingly being used for last-mile delivery services, where small delivery vans and trucks are required to navigate urban environments. These vehicles run entirely on electric power, which significantly reduces emissions compared to diesel- or gasoline-powered vehicles. They are particularly effective in cities with stringent emissions regulations and in regions where environmental concerns are heightened.

Hybrid Vehicles (HEVs):

Hybrid vehicles, which combine electric power with traditional fuel engines, offer flexibility in long-haul and intercity logistics. While not entirely emission-free, they deliver substantial reductions in fuel consumption and greenhouse gas emissions. Hybrid vehicles are especially advantageous for logistics companies that must operate over longer distances while transitioning towards a full electric fleet.

Benefits:

- Reduced carbon emissions and air pollution.
- Lower operating costs due to fuel savings.
- Compliance with increasingly stringent emissions regulations.
- Enhanced corporate social responsibility (CSR) and sustainability reputation.

Challenges:

- Higher upfront capital costs.
- Limited range and charging infrastructure for electric vehicles.
- Need for driver retraining and fleet adaptation.

2. Alternative Fuels: A Pathway to Decarbonization

While electric vehicles play a pivotal role in reducing emissions, alternative fuels such as hydrogen, biofuels, and natural gas offer additional opportunities to decarbonize logistics transportation, particularly for heavy-duty trucks and long-haul operations.

Hydrogen:

Hydrogen fuel cells emit only water vapor, making them one of the cleanest energy sources available. Hydrogen-powered trucks are being developed to replace diesel-powered long-haul trucks. Hydrogen has the advantage of a faster refueling time compared to electric vehicles and can be more suitable for heavy-duty transport where long range and high power are necessary.

Biofuels:

Biofuels, derived from renewable organic materials like plant oils or waste, can be used in existing internal combustion engines with minimal modification. Biofuels help reduce the dependence on fossil fuels and can lower lifecycle emissions by displacing petroleum-based fuels. Biofuels are particularly effective in regions where renewable feedstocks are abundant.

Liquefied Natural Gas (LNG):

Although still a fossil fuel, LNG is a cleaner alternative to diesel, emitting fewer particulates and lower CO_2 levels. LNG-powered trucks are increasingly being used in regions where natural gas infrastructure is available.

Benefits:

- Significant reduction in CO_2 and particulate emissions.
- Scalability in heavy-duty and long-haul logistics.
- Potential cost savings through alternative fuels.
- Diversification of energy sources away from fossil fuels.

Challenges:

- Infrastructure limitations, particularly for hydrogen and LNG.
- Potential competition with food and land use for biofuel production.
- The high cost of hydrogen production and fuel cell vehicles.

3. Automation and Artificial Intelligence (AI):

Optimizing Operations

Automation technologies, including robotics, drones, and autonomous vehicles, have the potential to revolutionize green logistics by optimizing resource use and reducing waste. Furthermore, AI and machine learning are playing a crucial role in improving logistics efficiency through data-driven decision-making.

Automated Warehouses:

Automated warehouses utilize robotic systems to handle goods more efficiently, reducing the need for energy-intensive human labor and optimizing space utilization. These robots can also minimize waste and errors, contributing to more sustainable operations. The integration of AI into warehouse management systems ensures that products are stored, retrieved, and dispatched in the most energy-efficient manner.

Autonomous Vehicles and Drones:

Autonomous vehicles (AVs) and drones are increasingly being tested for use in logistics. Self-driving trucks can optimize routes, reduce fuel consumption, and improve safety by eliminating human error. Drones are being used for last-mile deliveries, reducing the need for delivery vans and cutting down on traffic congestion and emissions.

Benefits:

- Improved energy efficiency in warehouses and transportation.
- Reduction in human error, leading to fewer wasted resources.
- Enhanced route optimization, reducing fuel consumption.
- Greater accuracy and speed in delivery operations.

Challenges:

- High initial investment in automation infrastructure.
- Regulatory hurdles regarding autonomous vehicles and drones.
- Technological limitations in fully autonomous systems.

4. Energy-Efficient Infrastructure: Green Warehouses and Distribution Centers

Energy-efficient design and construction of logistics facilities, such as warehouses and distribution centers, are vital components of a green logistics strategy. Sustainable buildings can contribute significantly to reducing the environmental impact of logistics operations.

Renewable Energy Integration:

Many modern logistics facilities incorporate renewable energy sources, such as solar panels, wind turbines, or geothermal heating and cooling systems. By generating their own clean energy, companies can reduce reliance on grid power and lower their carbon footprint.

Energy-Efficient Building Materials:

Sustainable building practices, such as using energy-efficient insulation, low-energy lighting, and eco-friendly building materials, help minimize the environmental impact of logistics infrastructure. Additionally, adopting smart technologies, such as building energy management systems, enables real-time monitoring and optimization of energy consumption.

Benefits:

- Lower operational costs due to energy savings.
- Compliance with green building certifications (e.g., LEED).
- Positive environmental impact through renewable energy adoption.
- Enhanced public image as a sustainable company.

Challenges:

- High upfront capital investment.
- Long payback period for energy-efficient technologies.
- Limited availability of skilled labor for green construction projects.

5. Smart Logistics: The Role of IoT and Big Data

The Internet of Things (IoT) and Big Data analytics have the potential to transform logistics operations into more sustainable, data-driven systems. By collecting real-time data from vehicles, warehouses, and inventory, companies can optimize their logistics processes and reduce waste.

Route Optimization:

IoT sensors embedded in vehicles and cargo allow for real-time tracking and monitoring. Advanced analytics can be used to determine the most fuel-efficient routes, reducing fuel consumption and emissions. GPS and traffic data can help avoid congestion, reducing delays and improving delivery efficiency.

Predictive Maintenance:

IoT sensors in vehicles and machinery can monitor equipment health and predict when maintenance is required. This reduces the risk of breakdowns, extending the lifespan of vehicles and equipment, and minimizing unnecessary repairs and replacements, which can lead to a significant reduction in waste and resource consumption.

Benefits:

- Reduced fuel consumption and emissions through efficient route planning.
- Reduced operational downtime through predictive maintenance.
- Increased transparency and visibility in supply chains.
- Lower operating costs through improved resource management.

Challenges:

- High initial setup costs for IoT and data infrastructure.
- Data security concerns related to IoT devices.
- The complexity of integrating IoT systems with existing logistics infrastructure.

Conclusion

Sustainable technologies are revolutionizing green logistics, offering innovative solutions to the environmental challenges facing the logistics sector. From electric and hybrid vehicles to alternative fuels, AI, automation, and energy-efficient infrastructure, these technologies enable companies to reduce their carbon footprints while improving efficiency and reducing costs. However, the implementation of these technologies comes with challenges, including high initial investment, regulatory hurdles, and technological limitations.

Despite these challenges, the benefits of adopting sustainable technologies—both for the environment and for businesses are substantial. As the demand for more eco-friendly practices grows, these innovations are crucial for creating a more sustainable and resilient logistics sector. By embracing these technologies.

Case Study

Mahindra Logistics: Leading Green Logistics in India

Overview:

Mahindra Logistics, a leading third-party logistics (3PL) provider in India, has committed to reducing its environmental impact through the adoption of green logistics practices. As part of the Mahindra Group's sustainability vision, the company has implemented various eco-friendly initiatives across its operations, from transportation to warehousing.

Key Sustainable Initiatives:

1. **Electric Vehicles (EVs):** Mahindra Logistics has pioneered the use of electric vehicles for urban deliveries in India. The company partnered with Mahindra Electric, a part of the Mahindra Group, to introduce EVs into its fleet for last-mile delivery operations. These EVs are used in cities like Mumbai and Bengaluru, reducing carbon emissions and contributing to air quality improvement.

2. **Fleet Optimization and Alternative Fuels:** The company has optimized its vehicle fleet by incorporating compressed natural gas (CNG) trucks alongside traditional diesel vehicles. CNG trucks emit fewer pollutants and are more cost-effective in the long term compared to diesel-powered trucks.

3. **Energy-Efficient Warehousing:** Mahindra Logistics has adopted energy-efficient warehousing solutions, incorporating technologies like LED lighting, solar power, and smart building systems to minimize energy consumption in its facilities. The company's warehouses are designed to be eco-friendly and reduce their carbon footprint.

4. **Carbon Footprint Reduction through Digitalization:** Mahindra Logistics has implemented digitization and data analytics to optimize routes and improve supply chain efficiency. This helps reduce fuel consumption and overall emissions. The use of telematics enables real-time monitoring of vehicles, which assists in route planning and reducing idle times.

Outcomes:

- **Environmental Impact**: By using EVs for urban deliveries and incorporating alternative fuels, Mahindra Logistics has significantly reduced its CO2 emissions and improved air quality in cities. The company's use of solar energy and energy-efficient technologies in warehouses has further minimized its carbon footprint.

- **Operational Efficiency**: Fleet optimization, digital tools, and route planning have reduced fuel costs and improved delivery times, making the logistics process more efficient and cost-effective.

- **Corporate Social Responsibility**: Mahindra Logistics' green initiatives align with the Mahindra Group's **"Rise for Good"** philosophy, enhancing the company's CSR profile and reputation as a sustainability leader in India.

Challenges:

- **Infrastructure Limitations**: While Mahindra is making strides with EVs, charging infrastructure in India remains limited, particularly in rural or remote areas, which can hinder the broader adoption of electric delivery vehicles.
- **High Initial Investment**: The upfront cost of electric vehicles and energy-efficient warehouse systems is high, requiring significant investment and long-term planning.

Future Outlook:

Mahindra Logistics aims to expand its EV fleet and further integrate renewable energy into its operations. The company is also focused on enhancing its digital capabilities to make logistics even more efficient and sustainable, in line with its goal to reduce its carbon footprint further in the coming years.

Conclusion:

Mahindra Logistics exemplifies how Indian logistics companies can adopt sustainable practices to reduce their environmental impact. Through the use of electric vehicles, alternative fuels, energy-efficient warehouses, and digital technologies, Mahindra is leading the way in green logistics in India. Their efforts not only help the environment but also boost operational efficiency and strengthen their brand image as a sustainability-driven company.

References

https://www.sciencedirect.com/science/article/pii/S0959652622013456

https://link.springer.com/article/10.1007/s11356-023-25639-7

https://www.mdpi.com/2071-1050/16/22/9757

https://link.springer.com/

08

DRONES AND AUTONOMOUS VEHICLES IN LOGISTICS

About the Author – Mr. Yuvraj Singh is an MBA student specializing in Logistics and Supply Chain Management (MBA-LSCM) at Parul University, Vadodara. He completed his Bachelor's in Business Administration from Bhupal Nobles' University, Udaipur. Yuvraj's professional journey began with a marketing internship at Shankar Packaging Limited, where he gained valuable insights into how marketing strategies influence packaging and logistics processes.

He further honed his skills during his Advanced Summer Internship Program (ASIP) at OM Logistics, Pune, where he contributed to optimizing inventory management and supply chain operations. His time at OM Logistics deepened his understanding of the role technology plays in enhancing supply chain efficiency.

Yuvraj also has 2 years of professional experience as a project associate, where he was involved in content management, talent pipeline generation, company profiling & competitive mapping, and candidate database management. In this role, he successfully managed project timelines and budgets and coordinated with cross-functional teams, sharpening his leadership and problem-solving abilities.

Currently, Yuvraj is working on a comprehensive project titled "IoT in Logistics," exploring how Internet of Things (IoT) technology is revolutionizing the logistics industry by enhancing tracking, monitoring, and overall supply chain efficiency. Through his experiences and ongoing research, Yuvraj is committed to integrating technological advancements into logistics and supply chain management practices.

Introduction

The logistics and supply chain industry is undergoing a transformative shift with the integration of cutting-edge technologies like drones and autonomous vehicles (AVs). These advancements are revolutionizing traditional logistics operations, enabling faster deliveries, reducing operational costs, and minimizing environmental impact. With the growing complexity of global supply chains, these technologies offer innovative solutions to longstanding challenges. This chapter provides a comprehensive exploration of the transformative roles, operational benefits, and broader implications of drones and autonomous vehicles in logistics.

The Role of Drones in Logistics

1. Delivery and Last-Mile Logistics

Drones have emerged as a game-changer for last-mile delivery, often regarded as the most time-consuming and expensive phase of logistics. Their ability to bypass conventional transportation infrastructure enables swift and efficient delivery of goods. These aerial devices operate autonomously using advanced GPS, AI-driven navigation systems, and real-time data connectivity to identify optimal routes, avoid obstacles, and ensure timely delivery. The reduction in human involvement and reliance on road networks significantly enhances efficiency while minimizing environmental impact.

Example: Amazon Prime Air

Amazon's Prime Air program has pioneered the use of drones capable of delivering packages weighing up to 5 pounds within a 10-mile radius. Using GPS navigation, these drones reach customers faster than conventional delivery methods.

2. Inventory Management and Warehousing

In warehouse operations, drones are increasingly used for inventory monitoring and management. Equipped with high-resolution cameras and sensors, they provide real-time visual and data feeds. Drones can scan shelves, track inventory levels, and identify misplaced items without manual intervention. This automation reduces errors, accelerates stock audits, and enhances inventory accuracy. Integration with warehouse management systems allows seamless data synchronization and improved decision-making processes.

Example: JD.com Warehouses

JD.com uses drones to scan barcodes and track inventory in real time, streamlining stocktaking and minimizing human error in its warehouse operations.

3. Disaster Relief and Humanitarian Aid

Drones play a critical role in delivering supplies during emergencies. Their ability to traverse difficult terrains and avoid obstructions makes them indispensable for relief operations. In scenarios where traditional logistics infrastructure is disrupted, drones ensure timely transportation of essential goods. Their small size and agility allow precise delivery to areas that are otherwise inaccessible, significantly improving disaster response efficiency.

Example: Zipline in Africa

Zipline drones deliver medical supplies like vaccines and blood to remote areas in Rwanda and Ghana, addressing healthcare challenges in underserved regions.

Autonomous Vehicles in Logistics

1. Freight Transportation

Autonomous trucks are reshaping freight transportation by enabling continuous operations without the limitations of human drivers. These vehicles rely on sophisticated systems, including LiDAR, radar, cameras, and AI algorithms, to detect road conditions, navigate traffic, and optimize fuel efficiency. By automating route planning and cargo handling, autonomous trucks reduce downtime and operational expenses while enhancing delivery reliability and consistency.

Example: Tesla Semi

Tesla's semi truck leverages autonomous technology for long-haul logistics, reducing fuel consumption and driver fatigue while improving efficiency.

2. Last-Mile Delivery Robots

Last-mile delivery robots represent a groundbreaking solution for urban logistics. These compact, ground-based autonomous systems navigate complex urban environments, including sidewalks and pedestrian crossings, to deliver goods directly to customers' doorsteps. Their design emphasizes low energy consumption, precise navigation, and secure package handling. Real-time tracking systems provide customers with visibility into delivery progress, further improving user experience.

Example: Starship Technologies

Starship's delivery robots are widely used in urban centers to autonomously deliver groceries and parcels, offering a cost-effective alternative to traditional delivery services.

3. Autonomous Warehousing

In warehouse environments, autonomous vehicles such as robotic forklifts and automated guided vehicles (AGVs) are transforming material handling tasks. These systems use sensors, AI, and advanced mapping algorithms to navigate storage facilities, transport goods, and perform repetitive tasks with high accuracy. Autonomous warehousing eliminates human errors, reduces operational risks, and significantly improves throughput, paving the way for fully automated logistics centers.

Example: Ocado Smart Warehouses

Ocado's fulfillment centers utilize autonomous vehicles for item retrieval, packing, and transportation, showcasing how automation can enhance operational scalability and accuracy.

Benefits of Drones and Autonomous Vehicles

1. **Cost Reduction**: Advanced automation minimizes labor dependency, lowers operational costs, and reduces waste.
2. **Speed and Efficiency**: Autonomous systems ensure faster processing and delivery by eliminating delays associated with human-driven operations.
3. **Sustainability**: Energy-efficient designs and the reduced use of fossil fuels contribute to lowering the carbon footprint of logistics operations.
4. **Enhanced Safety**: Automation reduces exposure to hazardous tasks, improving worker safety and minimizing accidents.
5. **Scalability and Flexibility**: Autonomous technologies provide scalable solutions capable of adapting to varying logistics demands and complexities.

Challenges and Limitations

1. **Regulatory Hurdles**: Compliance with evolving regulations concerning airspace usage, road safety, and liability introduces significant complexities.
2. **Technical Constraints**: Current limitations in battery technology, payload capacity, and operational range restrict the scalability of autonomous solutions.
3. **High Initial Investment**: The adoption of drones and autonomous vehicles requires substantial capital investment in infrastructure, technology, and training.
4. **Public Acceptance and Ethical Concerns**: The introduction of autonomous systems raises societal concerns about job displacement, privacy, and safety.

5. **Cybersecurity Risks**: Ensuring the protection of autonomous systems from potential cyber threats is critical to their secure and reliable operation.

Future Trends

Integration with AI and IoT

The future of drones and autonomous vehicles lies in their seamless integration with AI and IoT ecosystems. AI enhances predictive analytics, enabling smarter route optimization and adaptive decision-making, while IoT facilitates real-time connectivity and coordination across logistics networks. This synergy promises unprecedented levels of operational efficiency and responsiveness.

Example: FedEx IoT Integration

FedEx employs IoT-enabled sensors to track shipments and optimize delivery routes in real time, showcasing how interconnected systems improve performance.

Urban Air Mobility (UAM)

Urban air mobility represents a visionary shift in logistics, introducing the concept of drone taxis and urban freight drones. These systems aim to decongest roads and provide faster intra-city deliveries. UAM solutions require robust air traffic management systems and advanced vehicle designs to ensure scalability and safety.

Example: Uber Elevate

Uber's aerial logistics project aims to integrate drone taxis for intra-city package delivery, reducing congestion in metropolitan areas.

Collaborative Robotics

The convergence of autonomous vehicles, drones, and human workers will lead to hybrid models of logistics operations. Collaborative robotics leverages the strengths of human ingenuity and machine precision, creating highly efficient and adaptive workflows that address complex logistics challenges.

Example: Amazon Robotics

Amazon's robotics integrate drones and AGVs in fulfillment centers, ensuring seamless collaboration between automated systems and human employees.

Sustainable Innovations

Future advancements will prioritize sustainable technologies, such as solar-powered drones and electric autonomous vehicles. These innovations will further reduce the environmental impact of logistics operations and align with global sustainability goals.

Example: Solar-Powered Drones

Solar-powered drones are being tested by companies like Alphabet's Wing to enhance energy efficiency and reduce carbon emissions.

Conclusion

Drones and autonomous vehicles are redefining the logistics landscape, introducing a paradigm shift in efficiency, reliability, and environmental responsibility. These technologies have proven their ability to address critical logistical challenges, from last-mile delivery to disaster response, by leveraging automation, AI, and data-driven systems. Despite regulatory, technical, and societal hurdles, their transformative potential continues to grow.

The future of logistics lies in the widespread integration of these technologies into global supply chains. Businesses that proactively adopt and innovate with drones and autonomous vehicles will not only streamline their operations but also gain a competitive edge in a rapidly evolving market. As sustainability becomes an increasingly urgent priority, these solutions will also play a pivotal role in reducing carbon footprints and advancing green logistics practices.

By embracing this technological revolution, the logistics industry can enhance service quality, optimize resource utilization, and achieve unprecedented scalability. The road ahead is filled with opportunities to reshape the way goods are transported and delivered, making the supply chain more adaptive, sustainable, and resilient to future challenges.

Books and Academic Papers:

1. Chopra, S., & Meindl, P. (2019). Supply Chain Management: Strategy, Planning, and Operation. Pearson.
2. Christopher, M. (2016). Logistics & Supply Chain Management. Pearson UK.
3. Montreuil, B. (2011). "Toward a Physical Internet: Meeting the Global Logistics Sustainability Grand Challenge." Logistics Research, 3(2–3), 71–87.
4. Goodchild, A. V., & Toy, J. (2018). "Delivery by Drone: An Evaluation of Unmanned Aerial Vehicle Technology in Reducing CO2 Emissions in the Delivery Service Industry." Transportation Research Part D: Transport and Environment, 61, 58-67.

Industry Reports:

1. World Economic Forum (2020). The Future of the Last-Mile Ecosystem: Transition Roadmaps for Public- and Private-Sector Players.
2. McKinsey & Company (2022). The Future of Autonomous Vehicles in Logistics.
3. Deloitte (2023). Smart Logistics: Transforming Supply Chains with AI and Automation.

Case Studies:

1. Amazon Prime Air – Overview of drone-based last-mile delivery.
2. Zipline (2022). Impact Report on Drone Delivery of Medical Supplies in Africa.
3. Tesla Semi & Starship Technologies – Official company reports on autonomous logistics.

09

CYBERSECURITY IN SUPPLY CHAIN NETWORK

About the Author: Mr. Adarsh Dubey is an MBA candidate specializing in logistics and supply chain management at Parul Institute Vadodara, with a foundational background in B. Com from MSU Vadodara. I have gained practical experience through an internship at OM LOGISTICS PVT. LTD., where I was involved in the daily operations of logistics and warehousing management. My role included overseeing critical processes such as loading and unloading, challan verification, managing pick-up locations, and utilizing Oracle software for operational efficiency. This hands-on experience allowed me to deepen my understanding of supply chain dynamics, enhance my analytical capabilities, and develop a comprehensive awareness of industry regulations. The exposure to real-world logistics operations has equipped me with the skills and knowledge necessary to excel in the logistics and supply chain management field.

Introduction

As we all know, we live in a digital age, and our reliance on technology has grown for a variety of reasons, including convenience, affordability, cost savings, and many more. However, this benefit has become a burden for society as a result of the growing incidence of some malpractice in the environment, such as data breaches, which have called into question an individual's, society's, or firm's privacy and security. As a result, we take precautions to defend ourselves from such threats, giving rise to the term cybersecurity. Supply chain networks now require cybersecurity as a necessary component to guarantee seamless operations, data integrity, and business continuity. Let's learn cybersecurity in terms of supply chain management.

"Supply chain cybersecurity is the process of identifying, mitigating, and managing cyber risks to ensure the confidentiality, integrity, and availability of data within the supply chain ecosystem.". (Sengun Yeniyurt, 2021). In supply chain networks, cybersecurity involves safeguarding the interconnected systems and data exchanged between suppliers, manufacturers, distributors, and customers.

Importance

To protect from data breaches, supply chains manage sensitive data, including financial, supplier, and consumer information. Data breaches brought on by cyberattacks may result in monetary losses, harm to one's reputation, and legal consequences.

To Protection Against Operational Disruptions and Counterfeiting Goods: Any supply chain component that is the target of a cyberattack may experience operational disruptions. Ransomware assaults, for instance, can stop manufacturing, inventory, or logistics systems, resulting in delays and financial loss. They can also help to secure systems that track and validate the authenticity of products. By doing this, the possibility of fake items getting into the supply chain is decreased.

To Mitigating Third-Party Risks and Safeguarding Intellectual Property (IP); Many outside partners and vendors are a part of the current supply chain. The network as a whole may become vulnerable if one partner has poor cybersecurity, which might have an adverse impact and potentially lead to the theft of valuable technologies and a competitive disadvantage.

To Strengthening End-to-End Visibility: Cybersecurity improves supply chain visibility and gives stakeholders confidence in the veracity of the information used to make decisions.

To follow the nation's regulatory compliances, stringent regulations like GDPR, NIST, and ISO 27001 require organizations to ensure cybersecurity compliance across their operations, including third-party vendors.

Major cybersecurity threats in supply chain networks

Ransomware attack: A ransomware attack is a type of cyberattack where an attacker encrypts a victim's computer system or data and demands a ransom in exchange for the decryption key. Example: The Colonial Pipeline ransomware attack (2021) is a notable incident where the supply chain was severely disrupted. Attackers deployed ransomware that halted fuel distribution along the East Coast of the

United States. Colonial Pipeline paid a ransom of $4.4 million to regain access, but the disruption caused widespread panic, fuel shortages, and financial losses across multiple industries. This attack highlighted the vulnerabilities of critical infrastructure within supply chain networks.

Phishing and social engineering: whenever the attackers try to influence or manipulate the internal workforce of the organization or any other supply chain network through some deceptive emails, phone calls, or some fake websites to gain unauthorized access to networks.

IoT and hardware exploits; Sensors, RFID tags, and connected machinery are examples of IoT devices that have become essential components of today's logistics operations. However, these gadgets frequently lack strong security safeguards, making them prime targets for attackers. Hackers can use vulnerabilities in IoT devices to gain unauthorized access to networks, disrupt operations, or install malicious software. Example Mirai Botnet Attack in 2016, where IoT devices like routers and cameras were compromised and used to launch a massive Distributed Denial of Service (DDoS) attack. This attack disrupted major internet platforms and services, demonstrating the scale and impact of IoT vulnerabilities.

Distributed denial of service attack; A Distributed Denial of Service (DDoS) attack is an illegal attempt to overwhelm a network, server, or system by overloading it with large amounts of traffic. This overflow stops lawful consumers from using services, affecting business operations and resulting in financial losses. DDoS attacks can have serious effects on supply chain networks, which rely on real-time data and system availability to function properly. Example In 2016, the Dyn DDoS attack severely disrupted global internet services. The attackers targeted Dyn, a major Domain Name System (DNS) provider, using the Mirai botnet—a malware that infected insecure IoT devices such as routers, cameras, and sensors. The botnet coordinated an enormous amount of traffic to overwhelm Dyn's servers, leading to downtime for popular services like Amazon, Twitter, and Netflix. The attack demonstrated how insecure IoT devices could be weaponized to launch large-scale disruptions, affecting entire supply chains that rely on cloud-based services and real-time data.

Third-Party Vulnerabilities: Third-party vulnerabilities arise when external suppliers, vendors, or partners within the supply chain have inadequate cybersecurity measures. Because they depend on outside parties for logistics, hardware, and software services, organizations are vulnerable to cyberattacks. In order to get access to the networks, data, or systems of the main organization, cybercriminals frequently target these weaker links. Example In 2013, the retail giant Target experienced a massive data breach affecting 40 million credit card records. Hackers infiltrated Target's network by compromising a third-party HVAC (Heating, Ventilation, and Air Conditioning) vendor. The attackers gained access to Target's internal systems through the vendor's poorly secured credentials. This breach highlighted the critical need for stringent third-party cybersecurity protocols and vendor management practices.

Steps to overcome the threat of cybersecurity

1. Develop a cybersecurity strategy using frameworks like NIST or ISO 27001 for guidance. (as per government standard)
2. Secure technology infrastructure by updating software regularly and using firewalls.
3. Use AI-powered monitoring technologies to detect unusual activity and instantly identify cyberthreats.
4. Third-Party Risk Management.
5. Educate and Train Employees and by training and development.
6. Follow legal and compliance adherence.

Case study

In 2017, Maersk, a global leader in shipping and logistics, was hit by the NotPetya ransomware attack. This cyberattack originated from compromised accounting software in Ukraine and spread across

Maersk's network, paralyzing operations in over 600 ports and terminals worldwide. The incident disrupted global trade, exposing vulnerabilities in interconnected supply chains.

Outcomes

- The adaptation of a zero-trust model, advanced threat detection, and blockchain technology transformed Maersk into a resilient and adaptive organization.
- Adopting technologies such as AI and blockchain can improve resilience.
- Real-time monitoring and a zero-trust approach are critical for threat detection and mitigation.
- Improving vendor management and third-party compliance is crucial for securing networked systems.

References

Bibliography

al, A. e. (2017). Understanding the mirai botnet.

Colonial Pipeline Company. . (2021). Retrieved from https://www.cybersecurityjournal.com

government, e. (n.d.). NIST. Retrieved from national institute of standard and technology: https://www.nist.gov

Merchant, Z. (2022, 0305). THE ECONOMIC STIMES OF INDIA. Retrieved from https://economictimes.indiatimes.com/tech/newsletters/ettech-unwrapped/notpetya-the-cyberattack-that-shook-the-world/articleshow/89997076.cms?from=mdr

Sengun Yeniyurt, S. C. (2021). Cyber Security and Supply Chain Management: Risks, Challenges, and Solutions. World Scientific Publishing Company.

WIKIPEDIA. (n.d.).

INTRODUCTION TO E-COMMERCE AND LAST-MILE DELIVERY INNOVATION

About the Author - Mr. Sunny Kumar Singh is an MBA candidate specializing in logistics and supply chain management at Parul University, Vadodara. He holds a bachelor's degree in botany from Maharaja Sayajirao Gaikwad University, Vadodara, where he laid the foundation for his academic journey. Sunny further honed his professional skills during his internship at Aayush Road Lines, gaining hands-on experience in logistics and transportation management.

Beyond academics, Sunny is a passionate and accomplished athlete. He has represented his state at the national level in volleyball, showcasing exceptional teamwork, leadership, and discipline. Additionally, he has competed at the district level in cricket, further demonstrating his versatile athletic prowess. Sunny's dedication to both academics and sports highlights his commitment to personal growth, resilience, and excellence.

With a blend of academic knowledge and practical experience, Sunny aspires to make a significant impact in the field of logistics and supply chain management while continuing to nurture his passion for sports.

1.1 Introduction

E-commerce has revolutionized the way businesses and consumers interact, offering a seamless and efficient platform for transactions over the internet. Since its inception, e-commerce has grown exponentially, reshaping traditional retail landscapes and enabling businesses of all sizes to reach global audiences. It provides the convenience of shopping from anywhere, anytime, with diverse payment options, personalized experiences, and vast product choices. The rise of mobile commerce, social commerce, and AI-driven recommendations has further solidified e-commerce as an integral part of modern life.

As e-commerce continues to expand, the importance of last-mile delivery has become increasingly apparent. This critical phase in the supply chain directly impacts customer satisfaction, as it determines the speed, reliability, and convenience of product delivery. Last-mile delivery innovation focuses on overcoming challenges such as traffic congestion, high operational costs, and environmental sustainability. By leveraging cutting-edge technologies like AI, IoT, and robotics, and employing creative logistics strategies, businesses are redefining how goods are delivered to the end user. This innovation is essential for enhancing customer experience while maintaining efficiency and profitability in the rapidly evolving e-commerce ecosystem.

1.2 Definition and Meaning

Definition: E-commerce, or electronic commerce, is the buying, selling, and exchanging of goods, services, or information over the internet. It includes various models such as B2C (business-to-consumer), B2B (business-to-business), and C2C (consumer-to-consumer).

E-commerce represents a significant shift from traditional retail, enabling businesses to operate digitally and consumers to access products and services conveniently from their homes or mobile devices. It integrates technologies like digital payments, mobile apps, and online marketplaces to provide a seamless shopping experience. The growth of e-commerce is fueled by advancements in internet accessibility, mobile technology, and consumer demand for faster, more efficient shopping solutions.

Meaning: Last-mile delivery innovation refers to advancements in technology, processes, and strategies aimed at improving the efficiency, speed, and cost-effectiveness of the final stage of product delivery, where goods are transported from a distribution hub to the customer's location.

The "last mile" is often the most complex and expensive part of the delivery process, significantly affecting customer satisfaction. Innovations such as autonomous delivery vehicles, drones, real-time tracking, route optimization, and eco-friendly logistics methods are transforming this segment. These developments aim to address urban congestion, reduce delivery times, and enhance sustainability, ensuring a better experience for consumers and a competitive edge for businesses.

1.3 Growth of E-Commerce and Its Impact on Logistics

E-commerce has experienced exponential growth in recent years, driven by increased internet penetration, the proliferation of smartphones, advancements in digital payment systems, and changing consumer preferences toward online shopping. The COVID-19 pandemic further accelerated this trend as lockdowns and social distancing measures pushed more people to adopt e-commerce for essential and nonessential purchases. Global e-commerce sales are projected to continue growing, fueled by innovations such as artificial intelligence (AI), augmented reality (AR), and personalized marketing.

This growth has also led to the rise of omnichannel retailing, where businesses integrate physical and online stores to provide a seamless customer experience. Key players like Amazon, Alibaba, and Shopify have significantly influenced how businesses and consumers engage in e-commerce, setting new benchmarks for efficiency and convenience.

The rapid expansion of e-commerce has transformed the logistics landscape in several ways:

Increased Demand for Last-Mile Delivery With the surge in online orders, logistics providers face higher volumes, requiring faster and more efficient last-mile delivery solutions. Companies have adopted technologies like drones, autonomous vehicles, and smart lockers to meet this demand.

Shift Towards Warehousing and Fulfillment Centers To meet the need for shorter delivery times, businesses are investing in strategically located warehouses and micro-fulfillment centers close to urban areas. These hubs enable same-day or next-day delivery services.

Integration of Technology Logistics operations now heavily rely on technologies like route optimization software, real-time tracking, and AI-powered demand forecasting. These tools improve efficiency, reduce costs, and enhance customer satisfaction.

Sustainability Challenges and Innovations The increase in delivery vehicles has raised concerns about carbon emissions. To address this, companies are adopting green logistics solutions, such as electric vehicles (EVs), eco-friendly packaging, and optimized delivery routes to reduce environmental impact.

Personalization and Flexibility Consumers now expect flexible delivery options, such as same-day delivery, pick-up points, and time-specific slots. Logistics providers have adapted to meet these personalized demands.

Global Reach and Cross-Border Logistics E-commerce has enabled businesses to sell internationally, increasing the complexity of logistics operations. Managing customs, tariffs, and international shipping has become crucial for businesses aiming to expand globally.

1.4 Challenges in Last Mile Delivery

Last-mile delivery, the final step in the supply chain, is a critical but complex process that directly impacts customer satisfaction. However, it presents several challenges for businesses striving to meet consumer expectations efficiently and cost-effectively:

1. High Costs

- The last mile accounts for a significant portion (up to 53%) of total shipping costs.
- Factors like multiple delivery attempts, fuel expenses, and labor costs make this stage expensive.

2. Urban Congestion

- Delivering in densely populated urban areas often results in delays due to traffic congestion.
- Limited parking spaces and restrictions in city centers further complicate operations.

3. Rural and Remote Deliveries

- Delivering to rural or remote areas is challenging due to long distances, low order density, and infrastructure limitations.
- This leads to higher costs per delivery.

4. Customer Expectations

- Consumers increasingly demand faster, flexible, and real-time delivery options, such as same-day or next-hour delivery.
- Meeting these expectations while maintaining profitability is a significant challenge.

5. Failed Deliveries

- Missed deliveries due to the absence of recipients lead to additional costs for re-attempts and delays.
- These also reduce efficiency and impact customer satisfaction.

6. Environmental Concerns

- Increased delivery volumes contribute to higher carbon emissions and urban pollution.
- Companies face pressure to adopt sustainable practices, which may require substantial investment in green logistics.

7. Scalability Issues

- Rapid spikes in demand, such as during holidays or sales events, can overwhelm logistics systems.
- Maintaining service quality during these periods is challenging.

8. Technology and Data Limitations

- Insufficient use of technology for route optimization, tracking, and fleet management leads to inefficiencies.
- Poor data integration between e-commerce platforms and logistics providers can result in delays and errors.

9. Security Risks

- Theft and loss of packages during the last mile are common issues, especially for high-value items.
- Ensuring package security while maintaining delivery speed is a balancing act.

10. Workforce Management

- Finding, training, and retaining skilled delivery personnel is a persistent challenge, especially in markets with high competition.
- Managing drivers and their schedules to ensure efficiency adds to operational complexity.

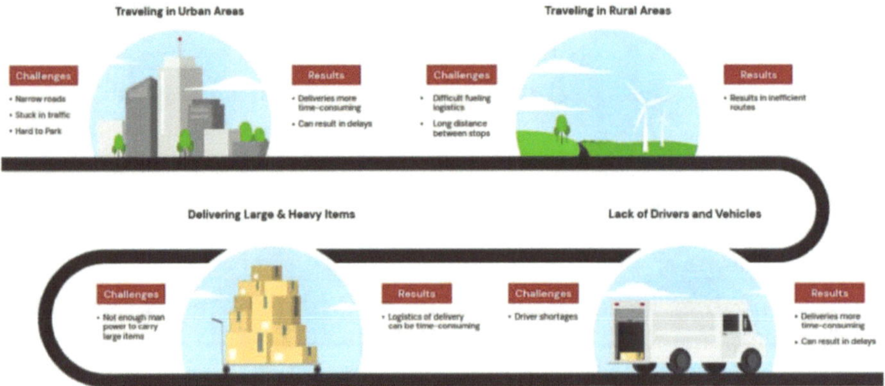

1.5 Technological Innovations in Delivery Systems

Technological advancements have significantly transformed delivery systems, enhancing efficiency, speed, and customer satisfaction. Below are key innovations that are reshaping the logistics and delivery landscape:

1. Autonomous Delivery Vehicles

- **Description:** Self-driving cars and trucks equipped with advanced sensors and AI navigate routes without human intervention.

- **Impact:** Reduces labor costs, improves safety, and ensures consistent deliveries, especially for long-haul logistics.

2. Drone Delivery

- **Description:** Unmanned aerial vehicles (UAVs) deliver packages, particularly in urban areas or hard-to-reach locations.
- **Impact:** Enables faster, contactless delivery while reducing traffic-related delays and emissions.
- **Examples:** Amazon Prime Air and Zipline.

3. Robotic Delivery Systems

- **Description:** Ground-based robots designed for short-distance deliveries, such as transporting goods from local hubs to customers.
- **Impact:** Enhances efficiency for urban areas and university campuses while reducing costs.

4. AI-Powered Route Optimization

- **Description:** Algorithms analyze real-time data (traffic, weather, road conditions) to identify the most efficient delivery routes.
- **Impact:** Minimizes fuel consumption, reduces delivery times, and lowers operational costs.

5. Smart Lockers and Pick-Up Points

- **Description:** Secure lockers located in public areas allow customers to pick up deliveries at their convenience.
- **Impact:** Reduces failed delivery attempts and enhances flexibility for customers.

6. Real-Time Tracking and Visibility

- **Description:** GPS-enabled systems provide live tracking updates for both businesses and customers.

- **Impact:** Improves transparency, enhances customer trust, and allows proactive management of delivery delays.

7. IoT (Internet of Things) Integration

- **Description:** IoT-enabled devices in delivery vehicles and packages provide real-time data on location, temperature, and security.
- **Impact:** Ensures the safe and efficient delivery of sensitive items like food, medicine, and electronics.

8. Electric and Sustainable Vehicles

- **Description:** The use of electric vehicles (EVs) and bicycles for eco-friendly delivery.
- **Impact:** Reduces carbon emissions and operational costs while promoting sustainable logistics.

9. Crowdsourced Delivery Platforms

- **Description:** Leveraging local individuals (gig workers) for deliveries through apps like Uber Eats, DoorDash, and Postmates.
- **Impact:** Expands delivery capacity and reduces overhead for businesses.

10. Warehouse Automation and Drones

- **Description:** Robots and drones in warehouses for picking, packing, and inventory management.
- **Impact:** Enhances order fulfillment speed and accuracy, optimizing the overall delivery process.

11. Blockchain for Secure Deliveries

- **Description:** Distributed ledger technology ensures secure tracking of packages, verifying delivery and reducing fraud.
- **Impact:** Increases transparency and builds trust in logistics operations.

12. Augmented Reality (AR) for Delivery Personnel

- **Description:** AR tools assist delivery drivers with navigation, package placement, and optimized routing.
- **Impact:** Improves delivery accuracy and reduces training time for new personnel.

1.6 Sustainability in Last Mile Logistics

Sustainability in last-mile logistics focuses on reducing the environmental and social impact of delivering goods to the end customer. As e-commerce and urbanization drive demand for faster and more frequent deliveries, the need for eco-friendly solutions has become paramount. Companies are adopting innovative practices and technologies to balance efficiency with environmental responsibility.

1. Adoption of Electric Vehicles (EVs)

- **Description:** Transitioning delivery fleets to electric vehicles reduces dependence on fossil fuels.
- **Benefits:** Significantly lowers carbon emissions and operating costs over time.
- **Examples:** Amazon's Rivian EV fleet and DHL's StreetScooter initiative.

2. Use of Cargo Bikes and Micro-Mobility Solutions

- **Description:** Employing bicycles and electric scooters for short-distance urban deliveries.
- **Benefits:** Reduces emissions, alleviates traffic congestion, and minimizes noise pollution.

3. Urban Consolidation Centers (UCCs)

- **Description:** Establishing localized hubs for consolidating packages before final delivery.
- **Benefits:** Decreases the number of delivery vehicles entering cities, optimizing routes and reducing emissions.

4. Smart Route Optimization

- **Description:** Using AI-powered tools to determine the most efficient delivery routes.
- **Benefits:** Reduces fuel consumption, delivery times, and overall environmental impact.

5. Eco-Friendly Packaging Solutions

- **Description:** Utilizing biodegradable, reusable, or minimal packaging materials.
- **Benefits:** Reduces waste and promotes recycling, aligning with customer preferences for green practices.

6. Crowd-Sourced and Shared Deliveries

- **Description:** Combining deliveries for multiple customers in a single trip through gig workers or shared platforms.
- **Benefits:** Reduces the number of trips and the overall carbon footprint.

7. Use of Drones for Delivery

- **Description:** Deploying drones for small package deliveries, particularly in remote or hard-to-reach areas.
- **Benefits:** Minimizes fuel consumption and reduces road congestion.

8. Smart Lockers and Pick-Up Points

- **Description:** Centralized delivery points where customers can collect their orders at their convenience.
- **Benefits:** Reduces failed deliveries and optimizes driver routes, cutting emissions.

9. Data-Driven Predictive Analytics

- **Description:** Using data to forecast demand and optimize delivery schedules.
- **Benefits:** Prevents over-resourcing and reduces wasted trips.

10. Renewable Energy for Warehouses

- **Description:** Powering distribution centers with solar, wind, or other renewable energy sources.
- **Benefits:** Cuts energy costs and lowers the carbon footprint of logistics operations.

Key Challenges in Achieving Sustainability

1. **High Carbon Emissions:** Increased delivery volumes result in a higher number of vehicles on the road, contributing to carbon emissions and air pollution.
2. **Congestion in Urban Areas:** Traffic congestion leads to longer delivery times, increased fuel consumption, and greater environmental impact.
3. **Excess Packaging Waste:** Many last-mile deliveries generate significant packaging waste, often non-recyclable.
4. **Failed Deliveries:** Re-attempting failed deliveries results in wasted resources and increased emissions.

Benefits of Sustainability in Last-Mile Logistics

- **Environmental Impact Reduction:** Significantly lowers emissions and waste.
- **Cost Savings:** Efficient routes, EVs, and reduced waste lead to lower operational costs.
- **Enhanced Brand Reputation:** Consumers increasingly prefer eco-conscious companies.
- **Regulatory Compliance:** Aligns with stricter government regulations on emissions and waste management.

Conclusion

The rise of e-commerce has revolutionized the global retail and logistics landscapes, driving unprecedented growth and transforming traditional business models. As consumer expectations for speed, convenience, and transparency increase, last-mile delivery systems have become a critical focus for businesses. However, this evolution presents challenges, including high costs, urban congestion, environmental concerns, and scalability issues.

Technological innovations such as autonomous vehicles, drones, AI-driven route optimization, and IoT-enabled tracking have significantly enhanced efficiency and customer satisfaction in delivery systems. Furthermore, sustainability in last-mile logistics has become imperative, with solutions like electric vehicles, cargo bikes, smart lockers, and eco-friendly packaging leading the way toward greener operations.Addressing these challenges and embracing innovation is essential for businesses to remain competitive in the fast-paced e-commerce ecosystem. A balanced approach that prioritizes technological advancements, operational efficiency, and sustainability will not only ensure better customer experiences but also contribute to environmental and economic well-being.

In conclusion, the interplay of e-commerce growth, last-mile delivery innovation, and sustainable practices is shaping the future of logistics. By adopting cutting-edge technologies and eco-friendly strategies, businesses can navigate the complexities of modern supply chains while building a resilient and customer-centric delivery network.

ARTIFICIAL INTELLIGENCE AND MACHINE LEARNING APPLICATIONS IN LOGISTICS AND SUPPLY CHAIN MANAGEMENT

About the Author - Gaurav Singh is an MBA student specializing in logistics and supply chain management at Parul Institute of Engineering Technology, Vadodara. With a background in BA from Glocal University, he has also spent three years in Navy NCC. During this time, he had the opportunity to participate in the All-India Yachting Regatta Camp at INWTC Chilka Lake, Orissa.

Gaurav demonstrated his leadership skills by leading 16 directorates from across India at the All-India Advanced Yachting Regatta Championship in Goa, where he was honored with the Best Cadet Award. Known for his precision and leadership, Gaurav continues to excel in his endeavors.

Making Logistics Smarter with AI and ML: A New Era in Supply Chain Efficiency

Introduction: The Changing Landscape of Logistics

Logistics has long been the backbone of global commerce, ensuring that goods are delivered from point A to point B, often over vast distances and through complex supply chains. In the past, this required extensive manpower, meticulous planning, and a lot of guesswork. However, with the advancement of artificial intelligence (AI) and machine learning (ML), the logistics industry is undergoing a revolutionary transformation. These technologies are not merely improving logistics; they are reinventing how goods move across the globe.

AI and ML are enabling logistics companies to optimize operations, predict and address issues before they arise, and create more efficient, cost-effective, and environmentally friendly supply chains. From automated warehouses to predictive maintenance for vehicles, AI and ML are making logistics smarter, faster, and more resilient. But how exactly are these technologies transforming logistics? Let's dive deeper into their real-world applications and explore the profound changes they're driving.

- **Real-Life Applications: Bringing AI and ML to Life**

  - **Route Optimization: Navigating the Path to Efficiency**

One of the most prominent applications of AI and ML in logistics is route optimization. For logistics companies, one of the most pressing concerns is getting goods to their destinations as quickly and cost-effectively as possible. This involves analyzing a range of variables, including traffic patterns, weather conditions, roadwork, and even accidents. Traditionally, this would have required human planners to manually adjust routes, a time-consuming and often imprecise process.

Now, AI-powered systems take over this task by continuously analyzing vast amounts of real-time data. Machine learning algorithms are able to process this information and predict the fastest, most efficient routes for delivery trucks. These systems are capable of factoring in everything from traffic jams to adverse weather, providing drivers with alternate routes that reduce delays. For instance, companies like FedEx and BlueDart leverage AI to ensure that packages are delivered on time, even in the face of unpredictable challenges. By optimizing routes in real time, they can reduce fuel consumption, decrease carbon emissions, and, ultimately, save money.

- ○ **Demand Forecasting: Predicting the Future of Consumer Needs**

Another game-changing application of AI and ML in logistics is demand forecasting. Imagine you're running a global e-commerce platform like Amazon, which caters to millions of customers around the world. With so many variables at play—seasonal trends, shifts in consumer behavior, promotional events, and unforeseen factors like natural disasters—forecasting demand accurately is a monumental task.

ML models are designed to handle this complexity by analyzing massive datasets that include past purchasing trends, seasonal patterns, and even external factors like social media activity. By doing so, they can predict what products will be in demand, when they will be in demand, and how much inventory will be required to meet that demand. This helps businesses like Amazon ensure that they stock their warehouses optimally, avoiding both shortages and overstocking. By aligning inventory with predicted demand, logistics companies can reduce waste and improve overall efficiency, making the supply chain more agile and responsive.

- ○ **Warehouse Automation: Robots in Action**

Gone are the days when warehouses were filled with people manually picking, packing, and sorting items. With the rise of AI and ML, many warehouses are becoming highly automated, with robots handling tasks that once required human labor. These robots are powered by

AI, enabling them to learn and improve their tasks over time. For example, robots can quickly scan and identify items, picking them off shelves with remarkable accuracy, packing them for shipment, and even sorting packages based on destination.

The advantages of warehouse automation are clear. Robots can work around the clock, speeding up the packing and sorting process while reducing errors and improving order accuracy. This allows human workers to focus on more complex tasks, such as quality control, decision-making, and customer service. In addition, AI-driven automation reduces operational costs and enhances the overall speed of the supply chain, contributing to faster deliveries and higher customer satisfaction.

- **Predictive Maintenance: Preventing Breakdowns Before They Happen**

Imagine a scenario in which a delivery truck breaks down on a busy highway, causing delays in the delivery of hundreds of packages. Not only does this lead to a poor customer experience, but it also results in significant costs for the logistics company, including repair fees, towing charges, and lost productivity.

With predictive maintenance, AI and ML algorithms can analyze data from vehicle sensors to detect potential issues before they lead to a breakdown. These systems monitor the performance of critical vehicle components, such as the engine, brakes, and tires, and predict when they might require maintenance or replacement. By addressing these issues in advance, companies can avoid costly repairs and minimize unplanned downtime. Predictive maintenance also helps to improve safety and extend the lifespan of vehicles, further reducing long-term costs.

- **Enhancing Customer Experience: AI-Powered Interactions**

The logistics industry is increasingly focused on improving customer experience, and AI plays a pivotal role in this transformation. AI-driven chatbots are now able to answer customer queries 24/7, providing real-time responses to questions about package status, delivery schedules, and product availability. These chatbots are designed to simulate human interactions, offering a personalized and efficient customer service experience.

In addition to chatbots, machine learning algorithms provide real-time tracking updates, allowing customers to track their orders with precision. AI-powered platforms can offer estimated delivery times, provide alerts about potential delays, and suggest alternative delivery options. This level of transparency not only enhances the customer experience but also helps logistics companies reduce the number of customer service inquiries and complaints.

- **The Bigger Picture: Sustainability and Cost Savings**

While AI and ML are primarily associated with making logistics smarter and faster, they also have a significant impact on sustainability and cost savings. As the world becomes increasingly concerned about climate change and environmental impact, logistics companies are under pressure to reduce their carbon footprints and operate more sustainably.

One of the key ways AI and ML contribute to sustainability is through route optimization. By calculating the most fuel-efficient routes, AI reduces the amount of fuel consumed by delivery vehicles, which directly lowers emissions. Additionally, AI can help companies minimize waste by improving inventory management, ensuring that products are stocked based on actual demand rather than overstocking, which can lead to unsold goods being discarded.

These efficiencies translate into cost savings for logistics companies. By reducing fuel consumption, optimizing routes, and streamlining inventory management, companies can cut operational costs and pass those savings on to consumers. At the same time, the reduced environmental impact helps to create a more sustainable and responsible supply chain.

- ### Challenges in Adopting AI and ML

Despite the many benefits, adopting AI and ML technologies is not without challenges. For many logistics companies, the cost of implementing these technologies can be prohibitively high. The infrastructure required to support AI-powered systems, such as high-performance computing, data storage, and specialized software, can be a significant financial investment.

In addition to cost, there is a need for skilled professionals to manage and maintain these systems. AI and ML require a high level of expertise in data science, machine learning, and software engineering, and finding qualified talent can be a challenge, particularly for smaller companies.

Data security is another major concern. As logistics companies collect and analyze vast amounts of sensitive data, they must ensure that this information is protected from cyber threats. Companies must invest in robust cybersecurity measures to prevent data breaches and protect customer privacy.

- ### Conclusion: A Future Driven by Intelligence

AI and ML are no longer futuristic concepts—they are already transforming the logistics industry in profound ways. By optimizing routes, forecasting demand, automating warehouses, and predicting maintenance needs, these technologies are making logistics smarter, faster, and more efficient. But the impact extends beyond operational efficiency; AI and ML are also contributing to a greener, more sustainable supply chain.

As these technologies continue to evolve, the possibilities for further improvements are limitless. From enhancing customer experience to reducing environmental impact, AI and ML are paving the way for a smarter, more connected world. Logistics companies that embrace these technologies will be better equipped to navigate the challenges of the future, delivering goods more efficiently, cost-effectively, and sustainably than ever before.

In the coming years, AI and ML will continue to shape the future of logistics, making supply chains more agile, responsive, and intelligent. The logistics industry is poised for a future where technology not only supports operations but drives innovation and growth on a global scale.

References

1. Books

- **Russell, S., & Norvig, P.** (2020). *Artificial Intelligence: A Modern Approach* (4th ed.). Pearson.

    ○ This book offers foundational knowledge on AI, which could be used for explaining the core concepts behind AI's use in logistics and other industries.

- **Chui, M., & Manyika, J.** (2018). *AI Superpowers: China, Silicon Valley, and the New World Order*. Hachette Books.

    ○ This can offer insights into the global impact of AI, including logistics.

2. Journal Articles

- **Liu, S., Zhang, L., & Li, J.** (2020). *Applications of Artificial Intelligence and Machine Learning in the Supply Chain Management: A Review. Computers & Industrial Engineering*, 150, 106851.

    ○ This paper reviews AI and ML applications across industries, with a focus on supply chain management.

- **Wang, Y., & Zhang, D.** (2021). *Machine Learning in Logistics: Opportunities, Challenges, and Future Trends. European Journal of Operational Research*, 290(2), 412-428.

    ○ A comprehensive paper on the growing role of machine learning in logistics.

3. Industry Reports

- **McKinsey & Company**. (2020). *Artificial Intelligence in Logistics: How it Transforms Supply Chain Efficiency*. McKinsey Insights.

- ○ McKinsey regularly publishes reports on AI in business sectors, and this would help back the points on route optimization, warehouse automation, and more.

- **Gartner**. (2021). *Supply Chain Management and AI: How to Leverage the Power of Machine Learning for Logistics and Demand Forecasting*. Gartner Research.
  - Provides research-driven insights on how AI and ML are shaping logistics and supply chains.

4. Conference Papers

- **Zhang, Z., Li, Z., & Wang, J.** (2019). *AI-Based Routing Optimization in Logistics: Algorithms and Applications*. In *Proceedings of the 2019 International Conference on Logistics and Supply Chain Management*. IEEE Xplore.

BLOCKCHAIN TECHNOLOGY FOR SUPPLY CHAIN TRANSPARENCY

About the Author - Tithi Patel is a dedicated and ambitious individual currently pursuing an MBA in logistics and supply chain management from Parul Institute of Engineering Technology, Vadodara. She holds a BBA degree from The Maharaja Shivajirao University, Vadodara.

Tithi's journey has been marked by her active involvement in the Navy NCC Vadodara, where she served for three years. Her passion and commitment led her to participate in the All-India Yachting Regatta Camp held at INWTC Chilka Lake, Orissa, and she also had the opportunity to perform in the All-India Advanced Yachting Regatta Championship at INWTC Goa.

Tithi is recognized for her perfectionism and strong leadership qualities.

What is blockchain? A Simplified Overview for Supply Chains

Blockchain is a technology that has revolutionized the way we think about data storage and transactions. At its core, blockchain is a digital ledger, a type of database, that securely records transactions in a transparent, tamper-proof, and decentralized manner. It allows information to be shared across a network of computers without the need for a centralized authority or intermediary. Imagine a shared notebook among multiple participants, where every transaction gets written down, and once added, it cannot be altered. This ensures that the records are accurate, immutable, and trusted by all participants.

To understand how this works, think of a supply chain. The process of moving goods from the manufacturer to the retailer or directly to the consumer can involve multiple parties—each recording their part of the transaction in various forms. Traditional methods of recording transactions—such as paper logs, databases, or even email chains—can be prone to errors, fraud, or tampering. Blockchain, however, records every transaction in a way that makes it nearly impossible to change after the fact. Once a record is made, it is verified by a network of computers, ensuring its authenticity and preventing anyone from manipulating the data.

The decentralization aspect of blockchain is also important. Traditional systems often rely on a single central authority, such as a bank or a government, to verify transactions. In contrast, blockchain distributes the data across a network of computers (called nodes), which all work together to verify the transaction. This decentralized structure makes it more secure and trustworthy since no single party can control or alter the data.

Blockchain's potential extends far beyond cryptocurrencies like Bitcoin; it is now being applied in various industries, including supply chains. The transparency, security, and reliability offered by blockchain technology are transforming how goods and services are tracked, from raw materials to final products.

Why Transparency Matters in Supply Chains

Transparency is crucial in modern supply chains for several reasons. As consumers become more conscious of the products they purchase, they want to know where these products come from, how they were made, and who made them. Whether it's verifying the authenticity of organic products, ensuring fair labor practices, or confirming the environmental impact of a product's production, transparency builds trust. Here's why transparency is especially important in supply chains:

1. **Authenticity**: Transparency ensures that products are exactly what they claim to be. Consumers can verify the origin of a product, whether it's a bottle of honey or a pair of designer shoes. This is vital in preventing fraud, especially in industries prone to counterfeiting.
2. **Ethical Practices**: Consumers are increasingly concerned about the ethical implications of their purchases. Supply chain transparency allows businesses to showcase that their products were made with fair labor practices and with minimal environmental harm.
3. **Reduced Fraud**: Counterfeit goods are a significant problem in many industries, from luxury items to pharmaceuticals. Blockchain's immutability and transparency make it easier to trace the origin and movement of goods, helping to prevent fraud and protect brand reputation.
4. **Customer Trust**: Consumers are more likely to buy from businesses that are open about where their products come from and how they're made. Supply chain transparency fosters consumer confidence, which in turn enhances brand loyalty and drives sales.

By ensuring transparency, businesses can not only meet consumer demands for ethical and authentic products but also improve their own operational efficiencies, reduce fraud, and increase consumer satisfaction.

Blockchain in Action: Use Cases for Transparency

Blockchain is particularly effective in addressing transparency challenges within supply chains. Here are some real-world examples of how blockchain is improving transparency in various industries:

Tracking the Journey of a Product

One of the most fascinating applications of blockchain in supply chains is tracking the journey of a product from its origin to its final destination. Consider the example of coffee beans. When you scan a QR code on a coffee packet, blockchain records every step of the coffee's journey—when and where the beans were harvested, how they were transported, and who roasted them. This digital trail allows consumers to trace the origins of their purchases, verifying that the product is authentic and that it was sourced ethically. This information is stored in a decentralized ledger, ensuring that it cannot be altered or erased.

Fighting Counterfeiting

Counterfeiting is a global problem, especially in industries dealing with high-value items like luxury goods, electronics, and pharmaceuticals. Blockchain provides a solution by recording the origins and movements of products in an immutable ledger. For example, a luxury handbag's journey from the manufacturer to the retailer can be tracked and verified using blockchain. Each step of its supply chain, from production to sale, is recorded in a decentralized ledger. This process makes it easy for consumers to confirm the authenticity of a product and protects manufacturers from counterfeit goods damaging their brand.

Ensuring Food Safety

Food safety is a major concern in supply chains, particularly in cases of contamination. Consider a situation where a batch of lettuce is found to be contaminated with E. coli. Without blockchain, tracing the contaminated lettuce back to its source could take days or even weeks, leading to widespread public health risks. However, with blockchain, the entire supply chain is tracked in real time. If contamination occurs, the source can be identified within minutes, allowing for quicker recalls and safer food consumption. This level of traceability improves consumer safety and reduces the risk of foodborne illnesses.

Verifying Sustainability Claims

Many companies now make bold claims about their commitment to sustainability, whether it's reducing carbon emissions or using eco-friendly materials. But how can consumers verify these claims? Blockchain ensures that sustainability pledges are backed by verifiable data. For example, a company claiming to use recycled materials in its products can prove this by showing the entire lifecycle of the materials—from sourcing to production. Blockchain tracks every step of production, ensuring that sustainability claims are not just marketing gimmicks but verifiable actions.

Key Benefits of Blockchain for Supply Chains

Blockchain technology offers several benefits that can greatly improve the efficiency and trustworthiness of supply chains. Some of the key advantages include:

1. **Trust**: Blockchain eliminates the need for intermediaries, allowing stakeholders to trust the data rather than each other. Because the data is verified by multiple participants in the network, it is more reliable.

2. **Security**: Blockchain uses advanced encryption techniques to ensure that the data is secure and tamper-proof. Once a record is added to the blockchain, it cannot be altered or erased, which reduces the risk of fraud and data manipulation.

3. **Efficiency**: By automating processes like tracking, documentation, and verification, blockchain reduces the need for manual oversight and minimizes errors. This leads to faster and more accurate operations.

4. **Consumer Confidence**: Transparency breeds consumer trust. By showing exactly where products come from and how they were made, blockchain enables companies to build stronger relationships with customers and improve brand loyalty.

Challenges in Implementing Blockchain

While blockchain has tremendous potential, its adoption comes with several challenges. These obstacles must be addressed for blockchain to be widely adopted in supply chains:

1. **Cost**: Implementing blockchain technology can be costly, especially for small and medium-sized businesses. The infrastructure required, including the setup of the blockchain network, software, and training, can be a significant financial investment.

2. **Complexity**: Blockchain technology is complex and requires specialized knowledge. Businesses may struggle to find professionals with the necessary skills to implement and manage blockchain systems, which can delay adoption.

3. **Integration Issues**: Many supply chains already rely on legacy systems that may not be compatible with blockchain. Integrating blockchain with existing systems can be difficult, requiring extensive adjustments to business processes and IT infrastructure.

4. **Data Privacy**: While blockchain offers transparency, it also presents challenges in terms of data privacy. Sensitive

information may need to be shared across a network, but businesses must balance transparency with the protection of personal or proprietary data.

The Future of Blockchain in Supply Chains

The future of blockchain in supply chains looks promising. As technology continues to evolve, blockchain is expected to become a mainstream tool for improving transparency, efficiency, and trust. Several developments are likely to shape the future of blockchain in supply chains:

1. **Global Standards**: As blockchain adoption grows, global regulations may be established to create uniform standards for its use in supply chains. This would enhance cross-border trade and streamline international logistics.

2. **Smart Contracts**: Smart contracts are self-executing contracts with predefined conditions. Blockchain will enable the automation of agreements, reducing the need for intermediaries and speeding up business transactions. Smart contracts could further streamline supply chain operations, from procurement to delivery.

3. **Widespread Adoption**: As the cost of implementing blockchain decreases and the technology becomes more accessible, small and medium-sized businesses will be able to adopt blockchain for their supply chains. This will make supply chains more transparent, efficient, and trustworthy on a global scale.

Conclusion: Building Trust, One Block at a Time

Blockchain is not just a technological breakthrough; it's a paradigm shift in how supply chains operate. By providing a secure, transparent, and tamper-proof system for recording and sharing data, blockchain addresses longstanding issues of trust, authenticity, and accountability in supply chains. As consumers demand more transparency, businesses that leverage blockchain will be well-positioned to meet these expectations and build stronger relationships with their customers.

Blockchain is paving the way for a more ethical, transparent, and efficient global supply chain. Whether it's verifying the origin of your food, ensuring the sustainability of your favorite brand, or confirming the authenticity of a luxury product, blockchain is redefining what it means to do business in the modern world. As this technology continues to evolve, its impact on supply chains will only grow, creating a more connected and accountable world for both businesses and consumers alike.

References

- **IBM**. (2020). *Blockchain for Supply Chain Transparency: A Case Study*. IBM Blockchain.

  - A case study illustrating IBM's blockchain solutions for supply chain transparency, particularly in industries such as food and luxury goods.

- **VeChain**. (2019). *VeChain Blockchain Solutions for Supply Chain Management: A Global Case Study*. VeChain.

  - A whitepaper detailing the use of VeChain's blockchain solution for enhancing traceability and transparency in global supply chains, particularly in industries like automotive and food.

- **Kavanagh, M.** (2020). *Blockchain in Supply Chain: How the Technology Is Transforming Logistics and Operations. Supply Chain Digital*.

  - An article discussing how blockchain is reshaping supply chain logistics, with examples from various industries.

- **Hodgson, G.** (2021). *Blockchain's Role in the Supply Chain: Enhancing Trust and Reducing Fraud. Forbes*.

  - A Forbes article detailing how blockchain technology reduces fraud, enhances transparency, and ensures the authenticity of products in supply chains.

- **ISO/IEC 20248** (2021). *Blockchain and Distributed Ledger Technologies for Supply Chain Management: Best Practices and Standards*.

  - Provides standards and guidelines for the use of blockchain technology in supply chain management, covering best practices for implementation and data sharing.

- **European Union Blockchain Observatory and Forum**. (2021). *Blockchain in the Supply Chain: Legal and Regulatory Challenges*.

  - A government report that addresses the legal, regulatory, and compliance challenges of implementing blockchain in supply chains across the EU.